The IBNET
Water Supply
and Sanitation
Blue Book 2014

The IBNET Water Supply and Sanitation Blue Book 2014

THE INTERNATIONAL BENCHMARKING NETWORK FOR WATER AND SANITATION UTILITIES DATABOOK

Alexander Danilenko

Caroline van den Berg

Berta Macheve

L. Joe Moffitt

WORLD BANK GROUP

ISBN (paper): 978-1-4648-0276-8
ISBN (electronic): 978-1-4648-0277-5
DOI: 10.1596/978-1-4648-0276-8

Cover photo: Alexander Danilenko
Cover design: Critical Stages

Library of Congress Cataloging-in-Publication Data has been applied for

Contents

BOXES

FIGURES

TABLES

Foreword

Well-run water utilities play an important role in ending poverty and boosting shared prosperity. Consumers need reliable access to high quality and affordable water supply and sanitation services. To deliver these basic services efficiently and effectively requires high-performing utilities that are able to respond to urban growth, connect the poor, and improve wastewater disposal practices.

Since 1997, the World Bank, through its International Benchmarking Network for Water and Sanitation Utilities (IBNET), has been working to improve utility performance through enhanced sharing of critical knowledge and expertise. An integrated part of the Water and Sanitation Program and new Water Global Practice, IBNET seeks to expand access to comparative data among utilities globally, helping to promote best practice among water supply and sanitation providers and eventually providing consumers with access to high quality and affordable water supply and sanitation services.

By delivering access to technical and financial information on utility performance, key stakeholders can do their jobs better: utility managers and employees can identify areas for improvement; governments can monitor and adjust sector policies and programs; regulators can ensure that customers get value; investors can identify viable markets and opportunities for creating value; and customer groups and NGOs can exercise "voice" in an informed way.

IBNET tools, such as data collection instruments and protocols, IBNET database, and IBNET tariff database, enable enhanced sharing of information from more than 4,000 utilities from over 130 countries and territories.

Due to its broad coverage and ease of accessibility, IBNET is becoming a standard for data collection and analysis around the world. More than 20 World Bank projects rely on IBNET as a reporting tool. Recently, Nigeria and Honduras water projects declared IBNET as an official monitoring tool for their utilities. IBNET also continues to support national and regional water associations. The Arab Countries Water Utilities Association (ACWUA) and the Pacific Water and Wastes Association (PWWA) recently benefited from IBNET adding their utilities to the network of peers, as well as providing information from their utilities' members. IBNET is also in use by the Swiss Development Agency and German Society for International Cooperation (GIZ). IBNET is growing its knowledge base of the sector. Its data collection module added an energy efficiency module that is becoming the standard for all new studies. The IBNET team is testing modules for assessment of the services for the poor and gender aspects in water and sanitation.

The IBNET *Blue Book* summarizes water sector development in 2006–11, describing trends and monitoring effects of recent crises. Despite difficulties, municipal water performance has improved and withstood accelerated urbanization and the impacts of the triple crisis (fuel, food, and finance). IBNET data confirmed that coverage with water services increased and water became accessible to more people. However, the financial crises of 2008 and 2010 significantly hampered the same trend with sanitation. As a result, wastewater services development has been slower than expected in many urban areas. The *Blue Book* also provides improved approaches in building comprehensive indices for water utilities performance assessment, as well as a statistical annex providing a comprehensive look at recent water performance of more than 80 countries.

We hope that this edition of the *Blue Book* will add to the knowledge of the sector and provide the reader with an informed view on municipal water and sanitation. We also hope that new utilities and their authorities will join IBNET and share their data with us and with the world.

Junaid Ahmad
Senior Director, Water Global Practice
World Bank Group

Jose Luis Irigoyen
Director, Transportation and ICT
World Bank Group

Acknowledgments

The preparation of the Report and its presentation was funded by the Water and Sanitation Program of the World Bank (WSP).

The main authors of the strategy are Alexander Danilenko, Caroline van den Berg, Berta Macheve, and L. Joe Moffitt. Key members of the team included WSP staff involved in water sector assessments in various countries, including, Dominick de Waal, Antonio Fernandez Serrano, Glenn Pearce-Oroz, Leila Elvas, Abdul Motaleb, David Michaud, and many others, without whom this report would have been impossible to produce.

We also thank the many national and international water associations, specifically, the Arab Countries Water Utilities Association (ACWUA), the Pacific Water and Waste Association (PWWA), and the Moldova Water and Wastewater Association (AMAC). We benefited significantly from the many water authorities and regulators who gave us valuable information on the water sector in their countries. Peer reviewers Manuel Marino, Luis Andres, William Kingdom, and Kirsten Hommann contributed valuable comments to this report. The team also valued the guidance and advice provided by TWI and WSP management, specifically from Jae So, Julia Bucknall, Bhuvan Bhatnagar, and William Rex.

About the Authors

Alexander Danilenko is a Task Team Leader for the International Benchmarking Network for Water and Sanitation Utilities (IBNET). Prior to this position, he was a team member of numerous World Bank projects and studies. He received his MSC in Resource Economics from University of Massachusetts (Amherst) and PhD in Physical (Environmental) Chemistry from Kazakh National Academy of Science.

Caroline van den Berg is a Lead Water and Sanitation Specialist in the Middle East and North Africa Region of the World Bank. She is the author of numerous publications on water and sanitation economics, sustainable water supply, and water services for the poor. Van den Berg holds an MSc in Economics and a PhD in Spatial Sciences from the University of Groningen.

Berta Macheve is an engineer who works on performance benchmarking of water and sanitation utilities in Africa, Asia, and Central and Latin America. Before joining the World Bank, she worked at the Mozambique Water Regulatory Council. She holds a MSc degree in Water Supply Engineering from UNESCO-IHE, Delft.

L. Joe Moffitt is Professor Emeritus in the Department of Resource Economics at the University of Masssachusetts, Amherst. He is the author of more than 50 publications focused on environmental and resource economics issues. Moffitt received his PhD in Agricultural and Resource Economics at the University of California, Berkeley.

IBNET Partners

ACWUA Arabic Countries Water Utilities Association
AMAC Moldova Water and Wastewater Utilities Association
ARA Romanian Water Association
CRA Mozambique Water Regulator
DWP Danube Water Program
EBC European Benchmarking Cooperation, the Netherlands
IUE Moscow Institute of Urban Economics, Russia
OECD Organisation for Economic Co-operation and Development
PWWA Pacific Water and Waste Association
SEAWUN South East Asia Water Utilities Network
SNIS Brazil National Information System for the Water Sector
WASREB Kenya National Water and Wastewater Regulator

Abbreviations

ACWUA	Arab Countries Water Utilities Association
AIC	Akaike information criterion
Apgar	Appearance, Pulse, Grimace, Activity, Respiration
GDP	gross domestic product
GIZ	German Society for International Cooperation
GNI	gross national income
IBNET	International Benchmarking Network for Water and Sanitation Utilities
IDB	Inter-American Development Bank
IWA	International Water Association
lcd	liters per capita per day
MDG	Millennial Development Goals
NRW	Nonrevenue water
O&M	Operation and maintenance
OCCR	Operating cost coverage ratio
PWWA	Pacific Water and Wastes Association
SAR	Special administrative region
WS	Water and sanitation
WSP	Water and Sanitation Program
WUVI	Water Utility Vulnerability Index

"Access to safe water is a fundamental human need
and therefore a basic human right."

—*Kofi Annan, United Nations Secretary General (2007).*

"Nothing is more useful than water: but it will purchase scarce anything;
scarce anything can be had in exchange for it."

—*Adam Smith, Chapter 4, Book I, Wealth of Nations (1776).*

OVERVIEW

The Success of IBNET

The International Benchmarking Network for Water and Sanitation Utilities (IBNET) has been involved in water sector monitoring since 1997. It collects data on utilities' performance and has set a global standard for performance assessment. As of December 2013, the IBNET database contained information from more than 4,400 utilities from 135 countries. This edition of the *Blue Book* sector assessment is based on intensive analysis and data collection for the period 2006 to 2011.

Water Sector Status

Safe and efficient water supply is a challenge for all countries and regions. For example, in newly established countries such as South Sudan, water service coverage from a protected source is virtually nonexistent (Nield 2011). In more advanced countries, such as the United States, "177 million US citizens currently get water from sources that lack adequate protection. Another 3.5 million are sick each year after exposure to bacteria found in raw sewage" (Cutright 2012). Virtually every report on water supply conditions speaks about the lack of proper maintenance or the negative effect of "reduced" maintenance efforts, the potential for reducing high electricity costs, and the large volumes of unaccounted-for water. Tariffs never satisfy utility needs and the public always considers them too high (regardless of the level); nonpayment or late payment is the norm for many consumers.

Nevertheless, governments, the private sector, and water authorities continue to devote substantial resources to develop the water sector. These ongoing efforts are an attempt to guarantee the right to water services that was declared as a human right by the UN in 2009. In this context, the performance assessment of water utilities serves the public interest of sharing prosperity by providing water and sanitation for all.

During the reviewed years of 2006 to 2011, municipal water performance has improved despite accelerated urbanization and the impacts of the triple crisis (fuel, food, and finance). Overall, water coverage increased and water became

1

accessible to more people. The financial crises of 2008 and 2010 significantly hampered the same trend with wastewater services, and now wastewater development cannot cope with urbanization in many places.

Achieving the Millennium Development Goals (MDGs) has been a major driver in the sector in the past decade. However, the financial crisis of 2008 had a major impact on the water and sanitation sector, as can be seen in water coverage rates. Water coverage declined in both low- and middle-income countries, but the decline was much more noticeable in low-income countries. The economic environment plays a significant role in water utility performance because all revenues come from the local constituencies and almost all costs are local.

At the same time, more and more utilities are actively handling water billing, collection, and water management through metering and other financial and technical operations. Metering is the norm for the vast majority of utilities. The proportion of utilities that could not cover their basic operation and maintenance (O&M) costs increased from 34 percent in 2000 to 37 percent in 2010. The effect is especially noticeable in low-income countries, where on average the percentage of utilities that could not cover even O&M costs increased from 28 percent in 2000 to 50 percent in 2010. Lower middle-income countries were the most affected, with 70 percent not able to cover their O&M costs. Upper middle-income countries seem less affected, partially because many continued to grow their economies rapidly; but even among these countries, 40 percent of water and sanitation utilities were not able to cover their basic O&M costs. Even in high-income countries, seven percent of utilities were still unable to cover their O&M costs in 2010.

Figure O.1   Cost Recovery of Water and Wastewater Services by Country Category, 2006–10

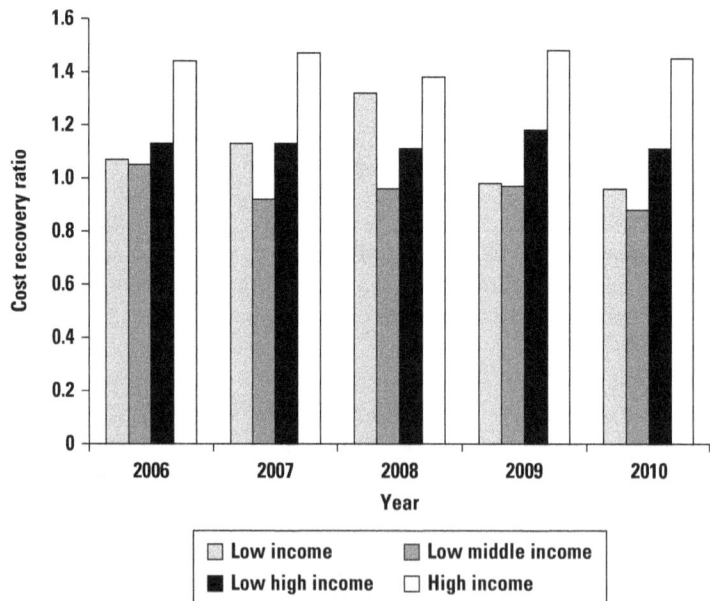

Source: IBNET database.

Most utilities, however, have passed on at least part of the growing O&M costs to consumers. Other indicators, such as median staff productivity and median collection periods, also show improvements. At the same time, the governmental role still dominates utilities' financial performance.

When analyzing the utility performance data, it should be taken into account that the economic status of many countries has changed rapidly. Average per capita nominal GDP increased rapidly from US$4,937 in 2006 to US$9,567 in 2011. The income classification in the IBNET sample has changed significantly as well, following rapid changes in economic development. As a result, many countries have moved into a higher income category: in 2000, 39 percent of utilities in the IBNET database were classified as from low-income countries, but only 6 percent were from this category in 2010.

The overall percentage of income spent on water continues to decrease as governments are reluctant to approve higher tariffs, but costs continue to rise. Median revenues have just kept pace with O&M costs—even though utilities became slightly more efficient as measured by improvements in staff productivity and nonrevenue water (NRW). At the same time, water has become even less affordable since the 2008 global financial crisis in many areas. Customers in low-income countries have substantially less ability to pay than customers in middle-income countries. In 2010, median affordability for households in low-income countries was 1.47 percent, 0.60 percent in high-income countries, and 0.86 percent in middle-income countries. Utility customers in low-income countries paid significantly more than those in middle-income countries, who in turn paid less than customers in high-income countries. This suggests that tariff structures currently being used in the sector to ensure that the poor get service are rather blunt instruments. In view of the rapid decline in affordability, there is scope to address tariff issues in a more constructive manner in both groups of middle-income countries, especially in higher middle-income countries.

Price increases negatively affect water consumption, whereas income growth positively impacts water consumption. Subsequently, water consumption shows diverse patterns between income groups and between regions, depending on economic growth trends and differences in real tariff developments. Median water consumption in liters per capita per day (lcd) stood at 158 liters in 2010. In the highest-performing quartile of utilities (defined as utilities with cost recovery of 1.30 and above), consumption stood at 118 lcd, compared to 218 lcd in the lowest-performing utilities quartile (defined as utilities with cost-recovery below 0.85) in 2010. At the same time, water consumption saw a sharp decline in low-income countries.

Improvement in technical operations became substantially harder to implement than originally thought. The rate of pipe bursts and unaccounted-for water did not change much in any group of utilities. This may be because most utilities have insufficient incentives for technical improvements, due to the nature of the operations and the cost structure of water and sanitation services. Our analysis suggests that there is no strong correlation between levels of NRW (as measured in cubic meters per kilometer per day) and economic development. On average, utilities in middle-income countries have poorer management and higher rates of NRW than in low-income countries. The median NRW in low-income countries was about 18 cubic meters per kilometer per day, whereas it was about 24 cubic meters in middle-income countries.

Aggregated IBNET Indices of Utility Performance: Apgar and WUVI

The definition of a successful water supply and sanitation service varies according to the observer. It ranges from "a human right and natural monopoly" to "a successful business that serves public welfare." Thus, water utilities are constantly caught in a difficult position and are subject to conflicting priorities. The utility must provide services to all customers at affordable prices while controlling quality and maintaining financial incentives for its staff. At the same time, it must control demand and thus reduce revenues. IBNET has continued piloting the concept of aggregated performance scores for utilities based on a variety of indicators.

The IBNET Apgar[1] score assesses a utility's health based on five indicators (six if the utility also provides sewerage services), which provide insight into the utility's operational, financial, and social performance. It was presented in the first edition of the IBNET *Blue Book*. We have devised a simple set of criteria that attempt to measure how utilities are doing overall, and not focus only on the financial and operational performance. These criteria are (i) water supply coverage, (ii) sewerage coverage, (iii) NRW, (iv) collection period, (vi) operating cost coverage ratio, and (vi) affordability of water and wastewater services. Each criterion is rated on a scale of 0 to 2, and then a total score is provided. For utilities that supply only water, the score is then normalized (as such utilities can only have a total score of 10 instead of 12). Using the overall Apgar score has also helped prevent overweighing the effects of general financial turmoil occurring around utilities.

The Water Utility Vulnerability Index (WUVI) is a dynamic version of Apgar. First, WUVI evaluates the utility's status by combining a few performance indicators into one consolidated measure. Second, it determines the threshold of efficient operation based on this consolidated index. Third, WUVI establishes a rating system based on the operational threshold that determines the probability of slipping into lower performance categories. The rating system also establishes the low performance rating at which municipal intervention is imminent.

The WUVI estimates the probability that a water utility will experience a performance problem as measured by its Apgar score in the future. The WUVI was developed for its predictive properties; it was conceived as a relational tool rather than a causal one, and it features significant predictors. Using the WUVI as an early warning device provides managers and policy makers an indication that further diagnostics are needed to determine the issues faced by a particular utility so that remedies can be put in place.

The subsequent chapters of this second edition of the *Blue Book* depict sector progress from 2006 to 2011. Chapter one discusses various aspects of sector performance from the viewpoint of each key performance parameter. Chapter two discusses how to define and measure a good utility. We revisit the IBNET Apgar score that was presented in the first edition of the *Blue Book* and suggest the addition of WUVI. This consolidated dynamic score can predict a water utility's future performance based on its current results. Annex 1 with IBNET countries' performance statistics will close this *Blue Book* edition.

Note

1. Appearance, Pulse, Grimace, Activity, Respiration. See http://en.wikipedia.org /wiki/Apgar_score.

References

Nield, R. 2011. "South Sudan Faces up to its Water Challenge." *Global Water Intelligence* 12(3): 8.

Cutright, E. 2012. "The Investment Drought." *Water Efficiency* 7(4): 8.

1

STATUS OF THE SECTOR

Introduction

The International Benchmarking Network for Water and Sanitation Utilities (IBNET) has been involved in water and sanitation sector monitoring since 1997. It collects data on utilities' performance and has set a global standard for performance assessment. As of December 2013, the IBNET database contained information from more than 4,000 utilities in 135 countries. In 2010 alone, the IBNET database reported performance results from 1,861 water utilities serving nearly 513 million people with water and 313 million with sewerage services in 12,480 cities and towns. This is approximately 14 percent of the total population of all households with piped water access in the world or nearly 45 percent of the urban population of developing countries. The database represents the equivalent of more than US$40 billion in annual revenue in 2010. The utilities represented in the database employ about 623,000 professional staff.

The *Blue Book* sector assessment is based on intensive analysis and data collection for the period from 2006 to 2011. Note that 2011 data are preliminary only as data collection is still ongoing. As the database has been growing rapidly, we will mainly focus on data from 2006 onward, which will ensure that the sample size per year is similar. The data from 2009 are the most complete in terms of all regions relatively well covered, with a total population of 664 million served with water and 331 million with sewerage services. Yet, the 2010 and 2011 data have sufficient depth to also be reported on. Information is collected from countries in all World Bank regions. However, data from countries in Africa and South Asia, which tend to be categorized as low income, are less-well represented for the year 2010 as data collection has not yet been completed. Hence, when reporting by income classification, data from 2009 are used for the analysis and reporting.

When analyzing the utility performance data, it should be taken into account that the economic status of many countries has changed rapidly. The nominal per capita GDP in the IBNET database increased rapidly from US$4,937 in 2006 to US$9,567 in 2011. The income classification in the IBNET sample has changed significantly as well, following rapid changes in economic development. Many countries have moved into a higher-income category. As a result, the number of low-income countries has decreased in favor of middle- and high-income countries. The number of utilities in low-income countries also decreased due to migration into higher-income groups. In 2000, 39 percent of the utilities in the IBNET database were classified as being located in low-income countries, while only 6 percent were from this category in 2010.

Data Quality

The quality of the IBNET database depends on the quality of the data submitted by individual utilities and utilities' associations. IBNET therefore invests substantial effort in making sure the data are of the highest possible quality and accurately and adequately reflect a reporter's performance. IBNET data come from a variety of sources, some of which have excellent quality assurance procedures (as in the case of regulatory data) and others of which follow less-sound procedures. To correct for this, IBNET continually improves its data-checking procedures and makes users aware of the quality (or lack of quality) of particular data. The need for rigorous quality assurance procedures is always balanced against the need to avoid discouraging potentially valuable data sources from participating.

Here we report median and unweighted average values of the performance indicators. However, the reader will note that many variables and performance indicators on which IBNET reports do not fit a normal distribution but rather form a skewed distribution. In these cases, a specific mean of the performance results distribution differs from its median. We regard the median as a better representation of performance than the average because of its skewness. As an additional effort to improve data quality, we exclude one percent of the outliers at both ends of the distribution of performance results.

IBNET Performance Reporting

Water Coverage (indicator 1.1)

Between 2000 and 2010, median water supply coverage expanded from 82 percent to 88 percent, despite the rapidly growing urban population. This can be attributed to the Millennial Development Goals (MDGs), which were a major driver of growth in the sector in the past decade. Water coverage varies with income level of the represented countries. As expected, utilities in low-income countries show lower water supply coverage rates than utilities in middle-income countries. In 2009, median water coverage for households in low-income countries was 62 percent, compared to 81 percent in lower middle-income countries, 93 percent in upper middle-income countries, and virtually 100 percent in high-income countries.

The 2008 financial crisis had a major impact on the sector, as can be seen in water coverage rates (table 1.1). By 2008, median water coverage had increased to 92 percent, but slipped to 88 percent by 2010. The decline in water coverage occurred in low- and middle-income countries.

In 2010, the best-performing quartile of utilities had water coverage rates of 98 percent or more, which practically amounted to universal coverage. The worst-performing quartile of utilities had water coverage rates of 69 percent or less.

Wastewater Coverage (indicator 1.2)

Median wastewater coverage[1] increased from 61 percent in 2000 to 76 percent in 2010. As table 1.2 shows, the number of utilities that provide wastewater services significantly increased. Nevertheless, wastewater coverage significantly lags behind water coverage. In addition, levels of wastewater coverage vary with economic development: utilities in low-income countries show lower wastewater coverage rates than utilities in middle-income countries. In 2010, average wastewater services coverage for households in low-income countries was 14 percent,[2] compared to 48 percent in lower middle-income countries, 77 percent in upper middle-income countries, and 89 percent in high-income countries. Most of the increase in wastewater coverage occurred in middle-income countries. Yet, since 2008, expansion has declined rapidly, suggesting that wastewater infrastructure investments are not keeping up with urban population growth in many places. The financial crises may also have affected investment volume in the sector.

In 2010, the best-performing quartile of utilities had wastewater coverage rates of 91 percent or more. The worst-performing quartile of utilities comprised those where wastewater coverage was 45 percent or less.

Nonrevenue Water (indicators 6.1 and 6.2)

There are a large number of indicators that measure nonrevenue water (NRW). The most common one is calculated as the difference between water produced and water sold, and measured as a percentage of water produced. Although very commonly used, it is not a very useful indicator as it is highly volatile. Other indicators measure NRW as the difference between water produced and water sold per kilometer of network, or per connection. Although one would assume that the different indicators are highly correlated, it can happen that a utility shows excellent performance in one NRW indicator but lesser performance in another. The International Water Association (IWA) has also mentioned that

Table 1.1   Coverage of Water Supply Services, 2006–11 (percent)

Indicator	2000	2006	2007	2008	2009	2010	2011 prelim
Median water coverage	82	92	92	92	90	88	89
Average water coverage	77	84	83	83	81	81	80
Standard deviation	23	20	21	20	21	21	22
Number of utilities reporting	630	1,454	1,534	1,507	1,725	1,686	1,453

Source: IBNET database.
Note: Prelim = preliminary. The 2011 data collection cycle is not yet complete.

Table 1.2   Coverage of Wastewater Services, 2006–11 (percent)

Indicator	2000	2006	2007	2008	2009	2010	2011 prelim
Median wastewater coverage	61	72	75	77	75	76	75
Average wastewater coverage	58	65	68	69	67	66	68
Standard deviation	33	28	28	28	29	29	28
Number of utilities reporting	438	957	1,031	993	1,069	1,144	1,028

Source: IBNET database.
Note: Prelim = preliminary. The 2011 data collection cycle is not yet complete.

data from one NRW indicator may give only partial information about the actual performance of a utility.

Median NRW (measured by the volume lost as a percentage of water production) declined from 31 percent in 2000 to 27 percent in 2011, while the standard deviation slightly increased. Hence, there are increasingly large differences in how utilities perform in this area (table 1.3).

Median NRW (measured as the volume lost in cubic meters per kilometer per day) has decreased from 26 cubic meters in 2000 to 17 in 2010. Yet, this indicator shows wide variations by year and between utilities (as can be seen in table 1.4).

In 2010, the best-performing quartile of utilities had a median NRW of 6 cubic meters per kilometer per day or less. The worst-performing quartile of utilities comprised utilities where this median indicator was 36 cubic meters or higher, as presented at the figure 1.1.

We found that interpretation of the NRW results is not necessarily straightforward. If one looks at NRW by income category, different NRW indicators can give a varied picture. For example, when measured by water volume lost as a percentage of water production, NRW is highest in low-income countries and lowest in high-income countries. Yet, when looking at other indicators as shown in table 1.5, different indicators do not always show the same trends.

Structurally reducing NRW has proven to be very difficult. Kingdom, Liemberger, and Marin (2006) mention that reducing NRW is not just a technical issue, but is also linked to weak management. When looking at a set of managerial indicators, the link between NRW performance and managerial performance is less than clear. We used a set of indicators to proxy for managerial performance, such as metering, collection period, staff productivity, average revenues per

Table 1.3   Nonrevenue Water as a Share of Water Production, 2006–11 (percent)

Indicator	2000	2006	2007	2008	2009	2010	2011 prelim
Median nonrevenue water	31	26	31	29	29	28	27
Average nonrevenue water	32	33	32	31	31	31	30
Standard deviation	17	16	17	17	17	17	17
Number of utilities reporting	589	1,242	1,448	1,349	1,403	1,488	1,253

Source: IBNET database.
Note: Prelim = preliminary. The 2011 data collection cycle is not yet complete.

Table 1.4   Nonrevenue Water, 2006–11 (cubic meters per kilometer per day)

Indicator	2000	2006	2007	2008	2009	2010	2011 prelim
Median nonrevenue water	26	20	21	18	19	15	17
Average nonrevenue water	41	34	32	29	30	27	28
Standard deviation	46	38	35	33	35	33	33
Number of utilities reporting	590	1,196	1,429	1,287	1,328	1,409	1,251
Highest-performing quartile	18	9	8	7	7	6	7
Lowest-performing quartile	68	45	44	40	41	36	37

Source: IBNET database.
Note: Prelim = preliminary. The 2011 data collection cycle is not yet complete.

Figure 1.1   Nonrevenue Water by Income Level: Median Values in 2010

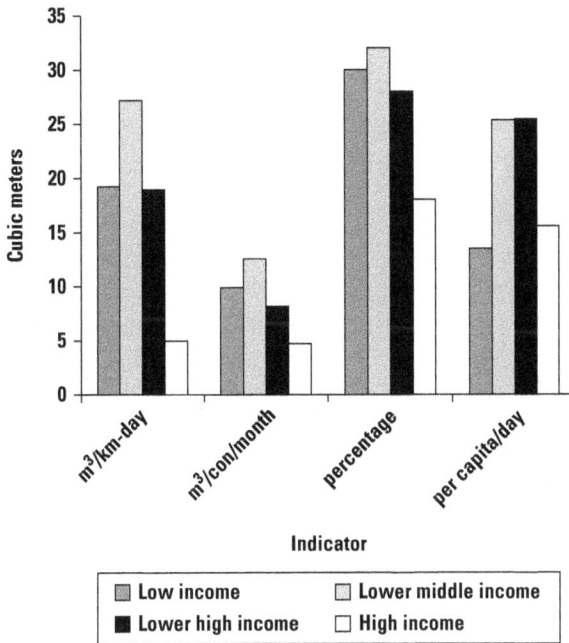

Source: IBNET database.
Note: m³/km-day = cubic meters per kilometer per day; m³/con-day = cubic meters per connection per day; m³/staff-day = cubic meters per staff member per day.

Table 1.5   Nonrevenue Water and Managerial Performance, 2006–11 (average values)

Level of NRW	Metering level (%)	Hours of supply per day	Collection period (days)	Staff productivity (staff per 1,000 connections)	Average revenues (US$ per cubic meter)	Operating cost coverage ratio
High	0.86	22	107	1.13	0.53	1.06
Average	0.99	22	76	1.00	0.68	1.14
Low	1.00	22	70	1.04	0.86	1.14

Source: IBNET database.
Note: NRW is measured in cubic meters per kilometer per day. The 2011 data collection cycle is not yet complete.

cubic meter of water sold, and operating cost coverage ratio. All of these indicators can be considered as proxies for the quality of utility management including collection efficiency, staff resource management, and financial management. As can be seen in table 1.5, it can be difficult to meaningfully compare utilities with low and high levels of NRW. The difference between most of these variables is statistically significant. Yet, it is not clear what the actual relationship is between the different variables, and what is the effect of the country environment on these variables. Higher average revenues per cubic meter of water sold provides utilities with more funds to reduce water losses, but average revenues per cubic meter of water sold also tends to be higher in high-income countries than in low-income countries. More research is needed to understand how far managerial performance and country environments affect NRW.

Table 1.5 also shows that the relationships between these different factors tend to be complex. Metering, which is often seen as a prerequisite for effective NRW management, tends to be an ineffective predictor for lower levels of NRW. However, a first analysis (see figure 1.2) suggests that widespread metering can be effective. Utilities with 100 percent metering show in general lower NRW than those with lower levels of metering. In this case, it might be surmised that high levels of metering produce benefits.

Staff Productivity

Staff productivity is measured as the number of staff per 1,000 connections, where higher productivity is reflected fewer staff per 1,000 connections. Staff productivity improved from 9 employees per 1,000 connections in 2000 to 7 employees in 2010. Yet, staff productivity varies widely from about 11 employees per 1,000 connections in low-income countries to slightly more than 3 in upper middle-income countries. This variance in staff productivity is partially linked to differences in connection practices. In many places in the world, water connections are shared among multiple households. Such an environment is often correlated with very low staff productivity. In Latin America, where most households have individual water connections, staff productivity is 3 staff per 1,000 connections. By contrast, in Eastern Europe and Central Asia, many apartment buildings are still fitted with a single connection; productivity was around 9 staff per 1,000 connections in 2009. In Africa, productivity in 2009 was about 10 staff per 1,000 connections, but other factors were involved. As African household surveys[3] increasingly show, many households are not directly connected to the piped network, but instead access (and often pay for) their neighbors' piped water.

Figure 1.2 Nonrevenue Water by Income Level: Median Values, 2006–11

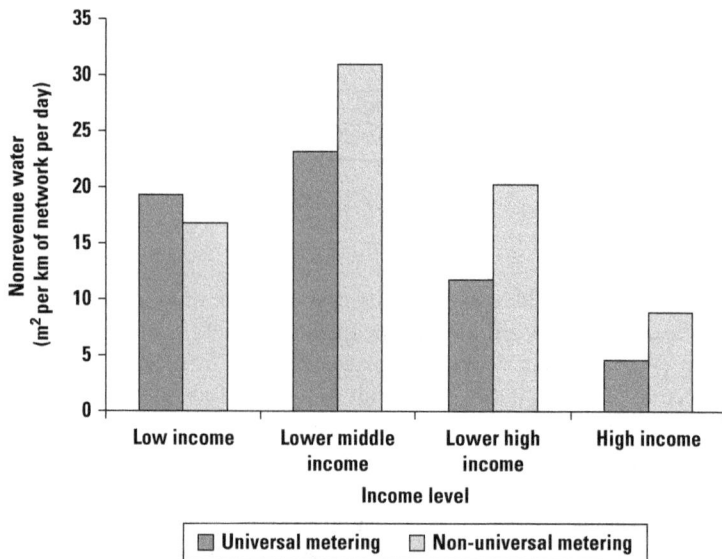

Source: IBNET database.
Note: Non-revenue water is measured in cubic meters per kilometer per day. The 2011 data collection cycle is not yet complete.

In 2010, the best-performing quartile of utilities had median staff productivity of 0.62 or less per 1,000 people served. The worst-performing quartile of utilities had median staff productivity of 1.57 or more. As table 1.6 shows, the gap between the best-and worst-performing quartile of utilities has decreased over time.

Part of the improvement in staff productivity could be the result of outsourcing staff functions. In such a case, the increase in staff productivity does not necessarily translate to lower staff costs. The staff cost trends differ significantly between regions.

Few high-income countries in the sample provide disaggregated details on their operating costs. Yet, data show that median labor costs as a percentage of total operating costs were about 36 percent in low-income countries in 2010, compared to about 40 percent in middle-income countries (table 1.7). The fast increase in median staff employee costs is directly linked to economic development, and to the fact that an increasing number of utilities participating in IBNET are now located in middle-income countries.

Operating Cost Coverage Ratio (indicator 24.1)

The median operating cost coverage ratio (OCCR) remained more or less constant between 2006 and 2010 at a level of 1.09 (table 1.8). The OCCR indicates that even in the best of times, the median utility barely covers its operating and maintenance (O&M) costs. Hence, it has no capacity to replace its assets once they wear out, let alone expand services to larger groups of consumers.

Unlike many other indicators, the difference between high- and low-performing utilities is relatively small for this indicator. In 2010, the high-performing utilities

Table 1.6 Median Staff Productivity Measured in Employees per 1,000 People Served, 2006–11

Indicator	2000	2006	2007	2008	2009	2010	2011 prelim
Median staff productivity	1.36	1.06	1.21	1.00	1.00	0.99	1.00
Average staff productivity	2.02	1.37	1.51	1.28	1.26	1.29	1.29
Standard deviation	1.77	1.09	1.22	1.04	0.98	1.03	1.01
Number of utilities reporting	598	1,421	888	1,440	1,679	1,803	1,574
Highest performing quartile	0.78	0.67	0.69	0.63	0.65	0.62	0.64
Lowest performing quartile	2.68	1.66	1.87	1.54	1.50	1.57	1.60

Source: IBNET database.
Note: Prelim = preliminary. The 2011 data collection cycle is not yet complete. The 2007 reporting rate was substantially lower than average.

Table 1.7 Median Staff Costs per Employee per Year, 2006–11 (US$)

Indicator	2000	2006	2007	2008	2009	2010	2011 prelim
Median staff employee cost	2,792	5,555	4,776	7,712	7,143	7,733	8,442
Average staff employee cost	4,568	7,170	7,346	9,438	8,959	9,751	10,538
Standard deviation	5,259	6,237	7,475	7,517	6,988	7,318	7,533
Number of utilities reporting	566	1,290	783	1,295	1,548	1,759	1,587

Source: IBNET database.
Note: Prelim = preliminary. The 2011 data collection cycle is not yet complete.

Table 1.8 Operating Cost Coverage Ratio, 2006–11

Indicator	2000	2006	2007	2008	2009	2010	2011 prelim
Median OCCR	1.10	1.09	1.11	1.10	1.15	1.09	1.09
Average OCCR	1.23	1.17	1.22	1.21	1.22	1.15	1.15
Standard deviation	0.50	0.52	0.58	0.56	0.59	0.54	0.58
Number of utilities reporting	565	1,447	1,420	1,494	1,449	1,664	1,429
Highest performing quartile	1.43	1.38	1.39	1.40	1.45	1.40	1.38
Lowest performing quartile	0.93	0.88	0.91	0.90	0.89	0.86	0.83

Source: IBNET database.
Note: OCCR = operating cost coverage ratio; Prelim = preliminary. The 2011 data collection cycle is not yet complete.

had a median value of 1.40 or more compared to the low performers with a median OCCR of 0.86 or less. Between 2006 and 2011, utilities in low-income countries registered a median OCCR of 1.09, compared to 0.99 in lower middle-income countries, 1.12 in upper middle-income countries, and 1.42 in high-income countries (table 1.8). However, while an OCCR of 1.4 means that the utility covers its O&M and depreciation costs, this is still far below the full cost recovery levels that utilities often aspire to. Hence, subsidies—whether in the form of investment and/or operating subsidies—remain a crucial resource for many utilities around the world.

The proportion of utilities that could not cover their basic O&M costs increased from 34 percent in 2000 to 37 percent in 2010. The effect is especially noticeable in low-income countries, where the share of utilities that could not cover even O&M costs increased from an average of 28 percent in 2000 to 50 percent in 2010. Lower middle-income countries were the most affected, with 70 percent not able to cover O&M costs. Upper middle-income countries seem less affected, partially because many of them continued to grow their economies rapidly. Nevertheless, 40 percent of the utilities in this quartile were unable to cover basic O&M costs. Even in high-income countries, 7 percent of utilities were unable to cover O&M costs in 2010.

As figure 1.3 shows, utilities in high-income countries generally have higher median OCCRs than low- and middle-income countries, but the difference is relatively small. Figure 1.3 shows that even a typical utility in a high-income country is not covering the full financial cost of the water and/or wastewater services. Utilities in lower middle-income countries tend to have the lowest median OCCRs. As countries get richer, the services provided also increase. In low-income countries, most utilities focus on delivering water supply services. In lower middle-income countries, wastewater collection is added, and in upper middle-income countries, a larger share of wastewater is treated. High-income countries have the most extensive wastewater services, and hence the cost of service increases, especially as wastewater collection and treatment tends to be relatively expensive.

Operation and Maintenance Costs (indicators 11.1 and 11.3)

Median O&M costs per cubic meter sold have increased rapidly since 2000, from US$0.28 to US$0.75 per cubic meter in 2010. The large standard deviations

Figure 1.3   Operating Cost Coverage Ratio by Income Category, 2006–10

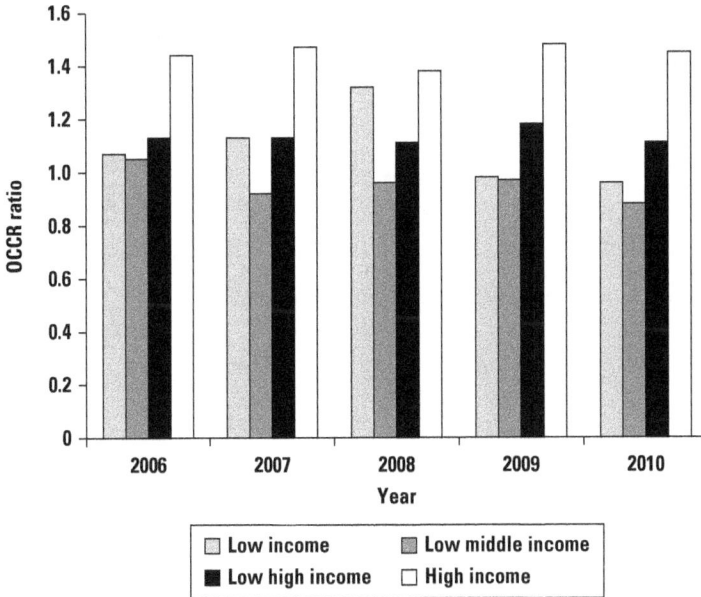

Source: IBNET database.
Note: OCCR = operating cost coverage ratio. The 2011 data collection cycle is not yet complete.

suggest a wide divergence between utilities in the cost of water and wastewater services.

There is also a wide variance between the O&M cost per cubic meter of water sold by income categories. In low-income countries, O&M costs per cubic meter of water sold increased to US$0.68 in 2010—compared to US$0.73 in lower middle-income countries, US$0.70 in upper middle-income countries, and US$1.69 in high-income countries. These cost differences are largely due to variations in service levels provided, but other factors also play a role, such as general price levels in the countries and exchange rate fluctuations.

What constitutes a high- or low-performing utility is harder to assess with the O&M indicators. It is generally assumed that the lower the cost, the more efficient the utility. However, it is not uncommon that utility costs are kept artificially low because certain expenditures are postponed (such as maintenance expenditures). Other expenditures may be only partially registered because a government agency pays for them. In addition, O&M costs vary with other factors, such as service level provided. The more wastewater that is collected and treated to increasingly high standards, the higher the O&M costs will be.

The difference between the highest and lowest quartile of utilities is shown in table 1.9. In 2010, the lowest-performing quartile of utilities had a median O&M cost per cubic meter of water sold of US$1.16; the highest-performing quartile had costs of US$0.44 or less.

Differences between better- and worse performing utilities show that variation shows a typical pattern. In low-income countries, the bottom 25 percent of utilities shows costs of US$0.72 per cubic meter of water sold, compared to only

Table 1.9 Operation and Maintenance Cost per Cubic Meter of Water Sold, 2006–11 (US$)

Indicator	2000	2006	2007	2008	2009	2010	2011 prelim
Median O&M cost	0.28	0.50	0.58	0.60	0.62	0.75	0.70
Average O&M Cost	0.36	0.65	0.75	0.80	0.79	0.88	0.82
Standard deviation	0.29	0.53	0.58	0.65	0.62	0.61	0.57
Number of utilities reporting	544	1,264	1,468	1,381	1,415	1,565	1,304
Highest performing quartile	0.14	0.28	0.32	0.33	0.34	0.44	0.40
Lowest performing quartile	0.87	0.87	0.99	1.03	1.05	1.16	1.12

Source: IBNET database.
Note: Prelim = preliminary. The 2011 data collection cycle is not yet complete.

Box 1.1 Drivers of Operation and Maintenance Costs

Many factors drive operation and maintenance (O&M) costs. The main factor, however, is labor costs. The higher the staff costs per employee and the lower staff productivity, the higher the O&M costs. A 10 percent increase in staff cost per employee results in a 4 percent cost increase. A 10 percent decrease in staff productivity (that is, an increase in the number of staff per 1,000 people) results in an O&M cost increase of about 6 percent.

The other main driver of O&M costs is per capita water production. A 10 percent increase in per capita water production results in an average reduction of O&M costs of about 6 percent—suggesting important economies of scale. Reducing nonrevenue water (NRW) does not have much effect on the bottom line

of a utility. A 10 percent reduction in NRW reduces O&M by only 0.4 percent. The provision of both water and sewerage services is also an important cost driver; it corresponds with increased O&M costs.

Metering increases the O&M costs, but the effect is very small. A 10 percent increase in metering results in a minimal increase in O&M costs. Age of the system (as measured by number of pipe breaks) has a minimal effect. An increase in pipe breaks by 10 percent increases O&M costs by 0.3 percent.

Economic growth has a significant effect. If gross national income (GNI) per capita increases, utility costs increase too. A 10 percent increase in GNI percent results in about a 3 percent O&M cost increase.

US$0.12 for the top 25 percent of utilities. In lower middle-income countries, the variance between the bottom and top 25 percent of utilities varies between US$1.20 and US$0.16. In upper middle-income countries, the range narrows to US$1.02 for the bottom and US$0.43 for the top, and in high-income countries the range further contracts to US$2.22 and US$1.18. Operation and maintenance costs converge, which is partly the result of water and wastewater services showing much less variation. In low-income countries, most utilities provide only water services. Once countries become richer, wastewater collection and varying degrees of wastewater treatment are added to the services that the utility provides.

The data show that the proportion of energy costs in the total O&M cost increased from 19 percent in 2000 to 21 percent in 2010.[4] There are large differences between utilities. In the lowest-performing quartile of utilities, electricity makes up less than 13 percent of O&M costs, compared to 34 percent in the highest-performing quartile of utilities. Utilities also show differences depending on their income status. In 2010, electricity made up 36 percent of O&M costs in low-income countries, compared to 22 percent in middle-income countries and 11 percent in high-income countries.

During the same period, the proportion of labor costs in the total O&M cost increased from 36 percent in 2006 to 41 percent in 2010. The combined share of energy and labor costs increased from 55 percent in 2006 to 62 percent in 2010, which likely crowded out other expenditures. This is a commonly observed pattern: when a crisis hits, the first budget casualty is spending on maintenance, which is often accompanied by a decline in the quality of services provided.

Operating Revenues (indicator 18.1)

Median revenues per cubic meter of water sold (as a proxy for tariffs) increased from US$0.34 in 2000 to US$0.81 in 2010. The increase in O&M costs as discussed previously was accompanied by a simultaneous increase in revenues, suggesting that utilities tried to raise their prices to keep covering growing O&M costs. The increase in average revenues has been especially pronounced in low-income countries, where median revenues per cubic meter of water sold grew from US$0.25 in 2000 to US$0.56 in 2010. In middle-income countries, median revenues per cubic meter of water sold increased from US$0.51 in 2000 to US$0.77 in 2010. In high-income countries, the same indicator stood at US$2.42 in 2010.

Table 1.10 shows the difference between the highest- and lowest-performing quartile of utilities. In 2010, the highest-performing quartile of utilities had median revenues per cubic meter of water sold of US$1.34 or more, compared US$0.41 or less for the lowest-performing quartile. In high-income countries, the lowest quartile of utilities registered median revenues per cubic meter of water sold of US$1.80 or less in 2010 compared to US$3.03 or more for the highest quartile. Yet, in upper middle-income countries, median revenues per cubic meter of water sold for the lowest-performing quartile of utilities were US$0.45 or less and US$1.22 or more for the highest quartile. In low- and lower middle-income countries, median revenues per cubic meter of water sold show much more diversity. In lower middle-income countries, median revenues per cubic meter of water sold for the lowest-performing quartile of utilities were US$0.23 or less and US$0.87 or more for the highest quartile in 2010. Utilities in the lowest-performing quartile in low-income countries registered median revenues of US$0.13 or less per cubic meter, compared to US$0.80 or more for those in highest quartile.

Water Consumption (indicator 4.1)

Price increases negatively affect water consumption, whereas income growth positively affects water consumption. Subsequently, water consumption shows diverse patterns between income groups and between regions, depending on economic growth trends and differences in real tariff developments. Median water consumption stood at 158 liters per capita per day (lcd) (table 1.11). In the highest-performing quartile consumption stood at 114 lcd, compared to 218 lcd in the lowest-performing quartile in 2010.

Overall, median water consumption increased marginally from 150 liters per capita per day in 2000 to 162 in 2010. Yet, different regions show different trends. In East Asia, Latin America, and the Middle East, water consumption increased

Table 1.10   Average Revenues per Cubic Meter of Water Sold, 2006–11 (US$)

Indicator	2000	2006	2007	2008	2009	2010	2011 prelim
Median average revenues	0.34	0.54	0.66	0.69	0.70	0.81	0.72
Mean average revenues	0.40	0.74	0.87	0.95	0.94	1.00	0.91
Standard deviation	0.30	0.64	0.74	0.88	0.82	0.81	0.75
Number of utilities reporting	572	1,290	1,498	1,400	1,480	1,567	1,299
Highest-performing quartile	0.56	0.97	1.15	1.19	1.23	1.34	1.23
Lowest-performing quartile	0.17	0.29	0.33	0.34	0.36	0.41	0.35

Source: IBNET database.
Note: Prelim = preliminary. The 2011 data collection cycle is not yet complete.

Table 1.11   Water Consumption, 2006–11 (liters per capita per day)

Indicator	2000	2006	2007	2008	2009	2010	2011 prelim
Median water consumption	151	150	150	150	146	158	162
Average water consumption	181	171	172	172	168	180	182
Standard deviation	109	95	97	97	99	103	95
Number of utilities reporting	586	1,206	1,449	1,379	1,416	1,482	1,268
Highest-performing quartile	112	110	110	110	107	114	119
Lowest-performing quartile	216	206	205	204	203	218	221

Source: IBNET database.
Note: Prelim = preliminary. The 2011 data collection cycle is not yet complete.

during the same period. In Eastern Europe and Central Asia, water consumption showed neither an increase nor a decrease. African and South Asian utilities showed a decline in consumption levels.

The combined effect of tariffs and consumption levels is measured in an indicator as total revenues per capita (in U.S. dollars). This indicator has increased rapidly over the years, from a median value of US$18 per capita in 2000 to US$45 in 2010. Yet, the standard deviation is large and also increasing rapidly, showing that the trends between utilities are divergent. In low-income countries in 2010, this indicator had a median value of around US$7 per year. In middle-income countries, the indicator almost doubled from 2006 to 2010 to US$43 per year. In high-income countries, median total annual per capita revenues were US$147 in 2010 (table 1.12). Interestingly, the largest change in consumer outlays for water has taken place in high-income countries.

Collection Period (indicator 23.1)

The median collection period decreased from 154 days in 2000 to 70 days in 2010. This is a rapid improvement, and means that the median utility is achieving the commonly used benchmark of a 90-day collection period. This improvement confirms that many more utilities have started active collection of their unpaid bills. Yet, table 1.13 shows the large difference between mean and average collection period and reflects the wide variation in performance between utilities in the efficiency with which they collect their billed revenues.

Table 1.12   Total Revenue per Capita per Year, 2006–11 (US$)

Indicator	2000	2006	2007	2008	2009	2010	2011 prelim
Median total per capita annual revenue	18	25	36	35	34	45	44
Average total per capita annual revenue	24	41	58	67	59	68	77
Standard deviation	20	57	84	109	93	101	143
Number of utilities reporting	616	1,432	1,515	1,474	1,599	1,634	1,383

Source: IBNET database.
Note: Prelim = preliminary. The 2011 data collection cycle is not yet complete.

Table 1.13   Collection Period in Number of Days, 2006–11

Indicator	2000	2006	2007	2008	2009	2010	2011 prelim
Median collection period	154	96	87	82	76	70	66
Average collection period	233	157	142	139	125	116	121
Standard deviation	239	181	158	175	154	153	160
Number of utilities reporting	474	1,179	1,244	1,174	1,237	1,155	975

Source: IBNET database.
Note: Prelim = preliminary. The 2011 data collection cycle is not yet complete.

Table 1.14   Affordability in Percentage of GNI, 2006–11

Indicator	2000	2006	2007	2008	2009	2010	2011 prelim
Median affordability	1.05	0.86	1.00	0.92	0.76	0.59	0.55
Average affordability	1.37	1.14	1.27	1.11	0.84	0.73	0.71
Standard deviation	1.03	1.01	1.00	0.89	0.73	0.59	0.59
Number of utilities reporting	594	1,437	1,521	1,476	1,600	1,633	1,383

Source: IBNET database.
Note: Prelim = preliminary. The 2011 data collection cycle is not yet complete.

Affordability of Water and Sewerage Services (indicator 19.1)

Affordability is considered a major challenge in many countries. Yet, in the IBNET sample, the median affordability (measured as average revenue per capita as a percentage of GNI per capita) turned out to have improved significantly from 1.05 percent in 2000 to 0.69 percent in 2010 (see table 1.14 The actual numbers are likely to be even smaller as most utilities serve urban populations, which tend to have higher average incomes than rural populations, whereas the GNI per capita is a national average. In general, as can be expected, water and wastewater services are more costly than provision of water supply services only. Average affordability in 2010 was 0.70 percent for households using both water and wastewater services, and 0.40 percent for those only using water supply services.

Consumers served by utilities in low-income countries spent more of their income on water and/or wastewater services than consumers from utilities in middle-income countries. In 2010, median affordability for households in low-income countries was 0.82 percent, compared to 0.55 percent in middle-income countries, and 0.79 percent in high-income countries. In general, affordability by income group seems to show a U-shape form, with utility consumers in low-income countries paying significantly more than consumers in middle-income countries. Yet, utility consumers in middle-income countries pay less than utility customers in high-income countries. The rapid increase in affordability values indicates a need to address tariff issues more constructively in middle-income countries, especially in higher middle-income countries.

Cross-Subsidies (indicator 21.1)

The operating cost coverage ratio discussed earlier shows that subsidies to utility services tend to be large and are provided in the form of investments and/or operating subsidies. Most of these subsidies are provided through direct government grants to utilities. But in some cases, the subsidies are (partially) provided through cross-subsidies, where certain consumer categories (mostly commercial and industrial water users) subsidize residential consumers. The IBNET database provides some details on the level of cross-subsidies within utilities, but the data is rather incomplete, as only a small number of utilities provide this type of data. Yet, as can be seen in table 1.15, in 2010 the median utility charged industrial users up to 1.98 times more per cubic meter of water than they charged residential users. Even more interesting is the large standard deviation, which shows that utilities display very different behaviors and that those cross-subsidies vary widely between utilities. Five percent of low-performing utilities registered cross-subsidies, with industrial users paying at least 15 times more than residential users. The increased use of cross-subsidies in utilities reporting the data suggest that when faced with the need for higher revenues, utilities have tried to reduce the impact on residential consumers by putting more of the burden on industrial water users.

The data also show a direct relationship between the level of cross-subsidies and the proportion of industrial water consumption in total water consumption. The median cross-subsidy rate is 1.98, meaning that industrial water and/or wastewater rates are twice that of residential rates. Yet, the average cross-subsidy rates are much higher and show the wide variation between utilities in using cross-subsidies as a tool.

If the level of cross-subsidies is limited to less than 1, then industrial water consumption makes up 30 percent of total water consumption. If the level of cross-subsidies is between 1 and 2, then industrial water consumption drops to 26 percent, whereas at a cross-subsidy level of more than 2, it drops to less than 14 percent. High levels of industrial water tariffs do not automatically translate into more revenues per cubic meter sold as industrial consumers react to tariffs

Table 1.15   Cross-Subsidy Levels, 2006–11 (Ratio of Industrial to Residential Tariff)

Indicator	2000	2006	2007	2008	2009	2010	2011 prelim
Median cross-subsidy rate	2.40	1.89	1.69	1.99	1.99	1.98	1.96
Average cross subsidy		4.22	3.24	3.98	4.02	4.21	3.80
Standard deviation	9.73	6.04	4.37	5.39	5.63	5.54	5.24
Number of utilities reporting	303	553	691	574	589	464	371

Source: IBNET database.
Note: Prelim = preliminary. The 2011 data collection cycle is not yet complete.

(just like other type of consumers). Hence, there seems to be an optimal level of relatively modest cross-subsidies (between 1 and 2) that optimizes the average revenues per cubic meter of water sold.

Conclusions

IBNET tools can monitor water sector development. More and more countries and their utilities are joining IBNET, making information about the sector open and available to all audiences. In 2010, IBNET covered about 43 percent of the urban population of developing countries. IBNET and its country-specific modifications are being used in about 30 countries, and more than 15 World Bank projects use IBNET for project development, monitoring, and supervision.

Since 2000, municipal water performance has improved despite accelerated urbanization and the impacts of the triple crises (fuel, food, and financial). Overall coverage increased, and piped water supply and wastewater services became accessible to more people. Yet, the financial crisis has hampered further progress since 2008, as coverage declined for both water supply and wastewater. Coverage has been unable keep up with population growth due to declining investment in the sector. At the same time, labor and energy's share in total costs have increased, suggesting that the decline in investment has been accompanied by delays in maintenance. Despite these setbacks, many operational and financial indicators have shown improvements. An increasing number of utilities are operating in a corporate manner, which involves actively handling water billing, collection, and metered water management. Metering has become the norm for a vast majority of the utilities, making water supply more businesslike and professional. Although sector performance has improved over the last decade, it is clear that these improvements can be easily derailed by economic developments such as the fuel crisis of 2007 and the financial crisis of 2008. Similarly, rapid economic growth can have a positive effect on utilities' performance.

Governments continue to be the dominant factor in utilities' performance. Although tariffs (measured as a proxy of average revenues per cubic meter) have increased over time, the increases were barely enough to cover the O&M costs of services. As a result, the operating cost coverage ratio has not shown any significant changes over the past decade. Although economic growth has been impressive over the last decade (resulting in a larger number of middle-income and high-income countries), the overall income spent on water continues to decline as governments are reluctant to approve tariff increases. As a result, water and/or wastewater services have become more affordable since the global financial crisis in 2008 as water continues to be a heavily controlled industry.

The IBNET data show that it is important to determine the drivers of performance of utilities. A first analysis shows that operation and maintenance costs are mainly driven by developments in labor costs (measured by the number of staff working in utilities and the staff cost per employee), the levels of water production, the delivery of wastewater services, and the income level in the country. Thus, drivers of O&M costs are mostly fixed for many utilities, and influencing these costs may be far less easy than perhaps is thought.

Finally, a quick analysis of the different indicators shows that one single indicator is not necessarily predictive of the performance of a utility.

Different indicators, for instance, can suggest very different performance levels for NRW. Hence, when analyzing the performance of a utility, it is important to look at a set of indicators and the context in which the utility is operating.

Notes

1. In this report, wastewater coverage refers to household connections to the wastewater network, not to the actual treatment or disposal of this wastewater.

2. This number should be interpreted carefully as the number of utilities in this category in 2011 was somewhat smaller.

3. Data from Demographic Health Surveys and Multiple Indicator Cluster Surveys underlie most of the data collected by the UNICEF-WHO Joint Monitoring Program, which measures progress toward the achievement of the MDGs for water supply and sanitation.

4. The data are still being collected, and as a result the number of observations is relatively small. Only data with a sufficient number of observations will be reported upon.

Reference

Kingdom, B., R. Liemberger, and P. Marin. 2006. "The Challenge of Reducing Non-Revenue Water (NRW) in Developing Countries—How The Private Sector Can Help: A Look at Performance-Based Service Contracting." Water Supply and Sanitation Sector Board discussion paper series, no. 8. Washington, DC: World Bank. Available at: http:// documents.worldbank.org/curated/en/2006/12/7531078/challenge-reducing-non -revenue-water-nrw-developing-countries-private-sector-can-help-look-performance -based-service-contracting.

2

DEFINITION OF THE GOOD UTILITY: IBNET APGAR AND WUVI

Introduction

The definition of a successful water supply and sanitation service varies according to the observer. The range of meanings includes defining it as a human right and natural monopoly to a successful business that serves the public welfare. Water utilities are constantly caught in a difficult position and subject to conflicting priorities. The utility must provide services to all customers at affordable prices while controlling quality and maintaining financial incentives for its staff. At the same time, it must control demand and thus reduce revenues.

Financial performance of a utility is estimated using a variety of indicators. These include, among many others, (i) affordability (reflected in the proportion of income spent on water services, the collection rate, and accounts receivable); (ii) exposure to external cost factors (such as financial markets and the price of electricity and chemicals, which usually depends on international prices); (iii) proper tariff-setting principles and timely tariff corrections; (iv) municipal development objectives that drive water utility development; and (v) cross-subsidies among different categories of users and social programs (including free or discounted water for the poor without proper compensation to the provider). Measuring financial performance gets even more complex if water quality is added to the assessment.

However, there is always demand for one uniform score to reflect a utility's success. All water utility stakeholders desire this one score—owners, regulators, financial institutions, customers, and even water utilities associations. The latter commonly give awards to their members for "best performance" without specifying what that actually means.

In this chapter we review experiences of determining a good performing company. Then we introduce the IBNET Apgar score[1] and its dynamic derivative, Water Utility Vulnerability Index (WUVI). WUVI first evaluates the utility's status by combining a few performance indicators into one consolidated measure. Second, it determines the threshold of efficient operation based on this consolidated index. Third, WUVI establishes a rating system based on the

---

Box 2.1   The AquaRating System

The Inter-American Development Bank (IDB), in cooperation with the International Water Association (IWA), is developing the AquaRating system, which assesses the performance of water and sanitation (WS) service providers in a comprehensive way. The system combines about 20 performance parameters with a data quality rating to produce one score. Besides an overall rating of the utility, AquaRating offers detailed assessments of its various rating areas (see figure B2.1.1), an assessment of the reliability of the information provided by the utility, and guidance to improve management practices. The pilot version of the system in Spanish is already completed and in test phase.

Figure B2.1.1   AquaRating Program Evaluation Criteria

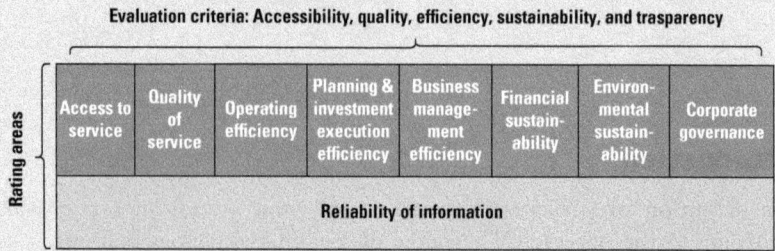

Evaluation criteria: Accessibility, quality, efficiency, sustainability, and trasparency

Rating areas	Access to service	Quality of service	Operating efficiency	Planning & investment execution efficiency	Business manage-ment efficiency	Financial sustain-ability	Environ-mental sustain-ability	Corporate governance
	Reliability of information							

Source: www.aquarating.org.

---

operational threshold that determines the probability of slipping into lower performance categories. The rating system also establishes the low performance rating at which municipal intervention is imminent.

The Search for a Good Scoring System

The definition of a good utility is always subjective and commonly based on the political context. Utilities at different development stages also defy a uniform performance definition. For example, in many countries, 24/7 service for 100 percent of customers is a norm for utilities; but this level of service is a distant dream for utilities in other countries. Most analyses of utilities' performance focus exclusively on financial and operational aspects. Besides these criteria, many countries assess the performance of utilities by how well they supply their service area population, including the poor.

Assessment gets even more complex when water quality is judged, because water is not a uniform product that can be easily compared.[2] Water quality standards and guidelines in Australia (HMRC, NRMMC 2011) are significantly different from those in Nigeria (SON 2007) or even in the EU (Council of the European Union 1998).

Many of these, and other attempts reported by the Pacific Water Association, Kenyan and Zambian regulators, the European Benchmarking Co-operation, and others to form a consolidated index, use 10–15 performance indicators. Unless properly weighted, use of such indicators in developing the aggregated score may lead to fluctuations in score components, making the final performance index less useful. In addition, weighting and normalization of indicators is often politically driven and biased to justify specific sector decisions or investment plans. As a result, the indicators can have little connection to actual performance.

The Apgar Score Revisited

IBNET has a unique opportunity to create a scoring system that uses our very large dataset to assess actual performance. The IBNET Apgar score assesses a utility's health based on five indicators (six if the utility also provides sewerage services), which provide insight into the utility's operational, financial, and social performance. It was presented in the first edition of the IBNET *Blue Book* (van den Berg and Danilenko 2011). We have devised a simple set of criteria that attempt to measure how utilities are doing overall, and not focus only on the financial and operational performance. These criteria are (i) water supply coverage, (ii) sewerage coverage, (iii) NRW, (iv) collection period, (vi) operating cost coverage ratio, and (vi) affordability of water and wastewater services. Each criterion is rated on a scale of 0 to 2, and then a total score is provided. For utilities that supply only water, the score is then normalized (as such utilities can only have a total score of 10 instead of 12).[3]

It should be noted that the Apgar score, particularly its set of indicators and the benchmark set, are based on characteristics of the IBNET database. It is quite likely that over time the Apgar score will be made up of different benchmarks and different indicators. As utilities develop, some indicators become less relevant, others more relevant. For example, in many developed countries, service coverage is almost universal and as such this indicator will likely be less important as a measure of performance. At the same time, benchmarks may change in value. When a sector shows improvement over time, the benchmarks need to be correspondingly adjusted for the Apgar score to remain relevant. Table 2.1 gives an overview of the Apgar score and its median values.

Each of these criteria is rated on a scale from 0 to 2, and then a total score is provided as the sum of the individual criteria scores. With six components, the score will be in the 0 to 12 range. Following the original Apgar scores, utilities are then classified as utilities that are *critically low* (with a score of 3.6 or less), fairly low (a score between 3.6 and 7), or normal (a score above 7). For utilities not providing wastewater services, the appropriate IBNET Apgar element is excluded and Apgar is calculated from 0 to 10. Table 2.2 shows the distribution of the reported indicators from the utilities based on information from the IBNET database.

Analysis revealed that about 90 percent of utilities that reached an Apgar score of 3.6 either experienced a transformation (went bankrupt and were subsequently renamed) or obtained a significant boost by compensating for accounts receivable, uncollected revenue, and delayed investments. An Apgar score of 5 pertains to utilities that barely cover O&M costs and are struggling with increased urbanization. Utilities with scores above Apgar 7 can be considered normal.

Table 2.1   Classification of Apgar Scores

	Indicator	Value	Average value of Apgar score for 2010
1.1	Water coverage	0 if < = 75%	
		1 if between 75% and 90%	1.14
		2 if > 90%	
2.1	Sewerage coverage	0 if < = 50%	
		1 if between 50% and 80%	1.20
		2 if > 80%	
6.2	Nonrevenue water	0 if > = 40	
		1 if > = 10 and < 40	1.09
		2 if < 10	
19.1	Affordability	0 if > 2.5%	
		1 if between 1.0% and 2.5%	1.78
		2 if < = 1.0%	
23.1	Collection period	0 if > = 180 days	
		1 if between 90 and 180 days	1.61
		2 if < 90 days	
24.1	Operating cost coverage	0 if < 1	
		1 if between 1 and 1.40	0.82
		2 if > = 1.40	
	Overall Apgar score	Critically low < = 3.6	
		3.6 < Fairly low < = 7.2	7.92
		Normal > 7.2	

Source: IBNET Database.
Note: The benchmarks as set reflect current database characteristics. The participation of ever more utilities and changes in their performance over time will likely necessitate adjustment of the benchmarks.

Application of the IBNET Apgar

The average IBNET Apgar score was 7.03 in 2011 (compared to 7.06 in 2008 when the Apgar score was first introduced). A trend of improvement was abruptly stopped in 2011, which reversed gains from 2000 to 2009. At the same time, a significant proportion of utilities had moved to a "normal" score by 2010. However, the 2010 financial crisis reversed these gains, as reflected in the 2011 IBNET Apgar breakdown. Still, the number of utilities with a performance classified as critically low has decreased compared with 2000, and even 2006 (see figure 2.1). This confirms the conclusions of chapter 1 of this report: the sector is growing and performance of water utilities is improving. However, improving trends are very vulnerable to the external economic environment.

Figure 2.2 shows the distribution of the utilities according to their IBNET Apgar in 2009. The form and shape of this figure is very similar for all years observed by IBNET.

Table 2.2   Apgar Score Value and Percentage of Each Category of Indicators in the IBNET Database

IBNET indicator	Apgar score value	Percent of observations
Water coverage (%)	0 if < 75%	34
	1 if > = 75% and < 90%	18
	2 if > = 90%	48
Sewerage coverage (%)	0 if < 50%	36
	1 if > = 50% and < 80%	27
	2 if > = 80%	36
Nonrevenue water (cubic meter per kilometer per day)	0 if > = 100	8
	1 if > = 40 and < 100	23
	2 if < 40	69
Collection period (days)	0 if > = 180	10
	1 if > = 90 and < 180	39
	2 if < 90	51
Affordability (water and wastewater bill as a percentage of GNI per capita)	0 if > = 2.5%	31
	1 if > = 1.0% and < 2.5%	26
	2 if < 1.0%	42
Operating cost coverage ratio	0 if < 1.0	37
	1 if > = 1.0 and < 1.40	40
	2 if > = 1.40	24
Overall Apgar score	Critically low < 3.6	20
	Low 3.6–5	29
	Fair 5–7	31
	Normal > 7	20

Source: IBNET database.

Figure 2.1   IBNET Apgar Score by Classification, 2000–11

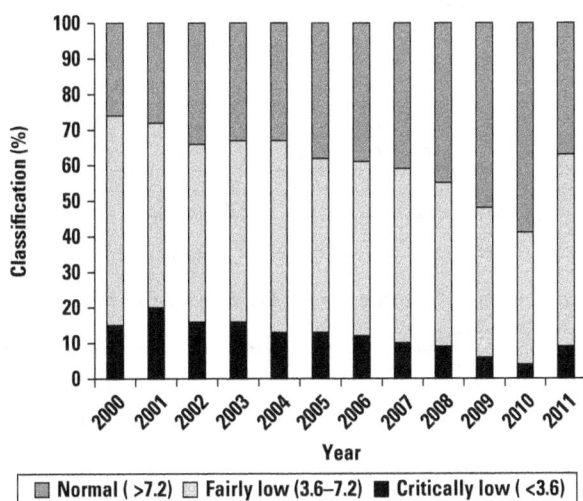

Source: IBNET database.
Note: The 2011 data collection cycle is not yet complete.

Apgar scores improved between 2000 and 2010, and the variance in performance between utilities slightly decreased, as measured by a decline in the standard deviation as shown in table 2.3. Thus, the performance pattern among them is getting closer.

Despite the positive trend in the performance of utilities, there are large differences in Apgar scores among utilities, individual countries, and groups of countries. Utilities in low-income and lower middle-income countries tend to have lower Apgar scores than utilities in higher middle-income countries. Moreover, table 2.4 shows that utilities in low-income and lower

Figure 2.2   Distribution of Utilities by Apgar Score, 2009

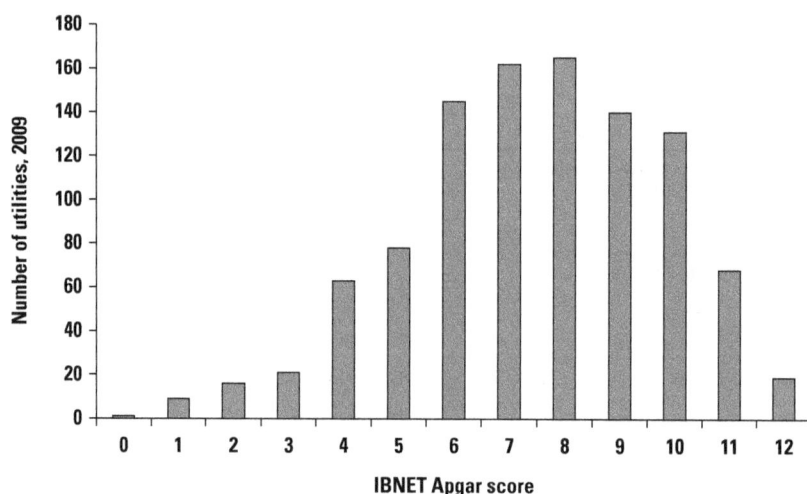

Source: IBNET database.

Table 2.3   Average Apgar Score, 2006–11

Year	2000	2006	2007	2008	2009	2010	2011 prelim
Apgar score	6.12	6.90	6.81	7.09	7.53	7.69	7.55
Standard deviation (US$)	2.30	2.31	2.17	2.16	2.23	2.03	2.00
Number of reporting utilities	526	935	1,119	983	997	1,006	891

Source: IBNET database.
Note: Prelim = preliminary. The 2011 data collection cycle is not yet complete.

Table 2.4   Unweighted Average Apgar Scores by Level of Economic Development, 2006–11

	2006	2007	2008	2009	2010	2011 prelim
Low-income countries	4.53	4.99	5.87	5.49	6.74	5.01
Lower middle-income countries	6.81	5.84	5.53	6.15	5.50	6.48
Upper middle-income countries	8.21	7.39	7.60	8.02	7.82	7.82
High-income countries	8.95	8.96	8.10	8.85	9.93	n/a

Source: IBNET database.
Note: Prelim = preliminary. The 2011 data collection cycle is not yet complete.

Figure 2.3 IBNET Apgar Score by Size of Utility, 2006–11

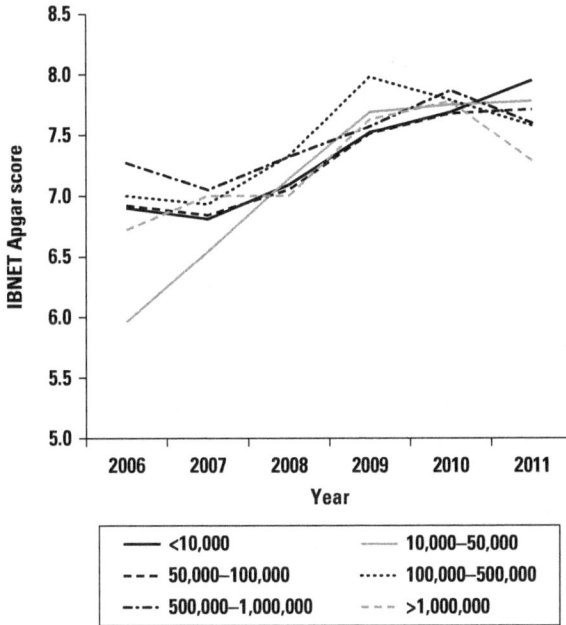

Source: IBNET database.
Note: The 2011 data collection cycle is not yet complete.

middle-income countries tend to be more vulnerable to external shocks than utilities in middle-income countries. The 2010 financial crisis had an effect on the Apgar score for utilities in low-income countries, but not for utilities in middle-income countries.

As can be seen in figure 2.3, size matters. Smaller utilities tend to have lower IBNET Apgar scores than larger utilities—up to a point. Very large utilities are not necessarily the most efficient.

Interestingly, smaller utilities have improved their performance more than larger utilities. Between 2006 and 2010, the new urbanized centers where utilities serve from 10,000 to 50,000 people improved the most, while utilities serving more than one million people saw the least improvement. Utilities that provide both water and sewerage services have a lower Apgar average un-weighted management score (6.99) than those that do not provide sewerage services (7.21). It is likely that more complex systems—that is, those that provide more than one service or provide services to large populations—require more management skills.

Water Utility Vulnerability Index (WUVI)

Definition

We designed WUVI as a dynamic version of the IBNET Apgar. WUVI in fact is an estimated probability that a water utility will experience a performance problem as measured by future Apgar score. First, we consider three different thresholds of

Apgar scores below which a utility is considered to be in a vulnerable position: an Apgar score of 3.6, a score of 5, and a score of 7. Hence, a WUVI depicts risk and the higher the threshold considered, the more "strict" the index becomes in the sense that the utility must have a high Apgar score to move out of the vulnerability zone. Second, we conceptualize a WUVI as an early warning device rather than an "actionable" index. By this, we mean that a high-value WUVI is a symptom of a possible future problem but does not indicate the specifics of that problem. Hence, we envision that managers and policy makers would treat a high-value WUVI as an indication that further diagnostics are desirable to determine the issues faced by a particular utility and to formulate potential remedies. From this perspective, the estimated WUVI is similar in character to many indicators already in use in other fields—most notably, the life sciences (Cabalu 2010; Cutter et al. 2010; Gnansounou 2008; Liang and Park 2010).

Third, and in a similar vein, our determination of a WUVI is relational rather than causal. We determine an association between current values of indicators and future water utility performance in order to best predict the likelihood of future performance in the critical range. Of course, this does not mean that the most closely associated indicator can be taken as the cause of a future problem. Such a determination would require diagnostic analysis focused on the underlying characteristics of the individual utility. Our methods detect statistical relationships that have, in the past, foreshadowed future water utility vulnerability regardless of whether or not the underlying complexity can be subjected to a detailed logical analysis. Such detection of associations for forecasting purposes is evident in fields ranging from security analysis to meteorology.

Following Estrella and Mishkin (1998), we use the probit analysis statistical technique with model selection to develop the WUVI. From a set of possible indicators, we found the weights and indicators that have been historically the most accurate in predicting a critically low future Apgar score. Again, following Estrella and Mishkin, our model uses a 0–1 indicator of a critically low Apgar score as a dependent variable. This is done because the risk of a critically low Apgar score is what the WUVI is meant to detect (rather than a point estimate of a specific utility's actual Apgar score). The details of the WUVI development were published in 2012 (Moffitt, Zirogiannis, and Danilenko 2012; Zirogiannis et al. 2012).

Properties of the WUVI Function

The most accurate prediction of water utility vulnerability two years in the future is achieved by constructing the WUVI with five variables evaluated at the current time. This specification of the WUVI minimized the Akaike information criterion (AIC) and was selected. The five variables included in the WUVI are (1) "Water Coverage," defined as the percentage of households in the utility's service area receiving water service from the utility; (2) "Sewer Coverage," defined as the percentage of households in the utility's service area receiving sewer service from the utility; (3) "Nonrevenue Water," defined as cubic meters per kilometer per day of water in the utility's service area for which the utility does not receive compensation; (4) "Affordability," defined as the utility's revenue as a percentage of per capita gross national income, and (5) "Collection Period," defined as the number of days required for the utility to collect payment for water and/or sewer services provided.

While it may be tempting to contemplate a causal explanation based on the predictive variables contained in or omitted from the WUVI, it must be kept in mind that the WUVI is relational in nature. Conceivably, any predictor or combination of predictors could and would have been used to construct the WUVI relationship, had such predictors most accurately foreshadowed future water utility vulnerability. Interpretation of the estimated WUVI in a causal, regression-type framework is incorrect, since such a model would almost certainly need to reflect substantial complexity in portraying the future vulnerability of a water utility. The focus in development of the WUVI was significance in prediction.

WUVI Examples

Moldova: Reforms and External Factors of Water Sector Development. The water sector in Moldova went through decentralization, painful tariff reforms, and demand management. Sector reforms resulted in significant reduction of the WUVI. This can be verified from the downward slope of the index in figure 2.4. The WUVI starts at around value of 75 percent in 1996 for utilities in both Chisinau and Balti. This suggests that there is more than 75 percent probability that in two years (that is, 1998) these utilities will have an Apgar score lower than 3.6. In other words, they are likely to be faced with major operational challenges and experience increased vulnerability. From 1996 onward, the WUVI value decreases, suggesting that the status of the utilities is improving. Figure 2.5 shows the corresponding Apgar values, which are stable and increasing during the last few years. The economic crises in 2000 and 2008 increased the vulnerability of the two utilities. The WUVI increases for both utilities during those two years, while the Apgar demonstrates a sharp decline.

Figure 2.4   WUVI Standard of the Two Largest Utilities in Moldova, 1996–2012

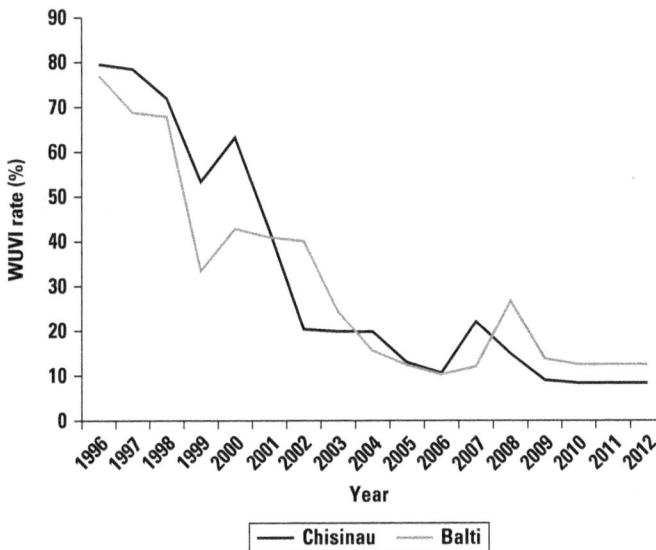

Source: IBNET database.

Figure 2.5 IBNET Apgar Score of the Two Largest Utilities in Moldova, 1996–2012

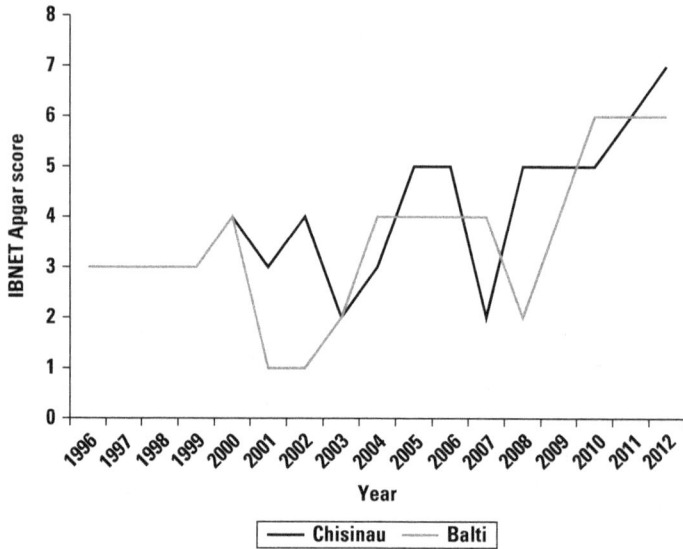

Source: IBNET database.

The main difference between the WUVI and the Apgar score is that the former provides information about the status of the utility two years into the future. The Apgar score, on the other hand, gives insight pertaining to current performance based on the six indicators.

Western Africa: Cost Recovery versus Vulnerability. Since the 1990s, national utilities in western African countries have been pursuing the goal of higher cost recovery of water and sanitation operations. Most utilities have achieved this goal, with some mixed results. In Gabon, despite significant tariff increases, water consumption stayed at the same level due to overall country development, thus keeping affordability relatively high. Figure 2.6 shows the value of the WUVI for Gabon, starting in 2001. That year the utility had a WUVI of 54 percent, suggesting that in two years there was a 54 percent chance of experiencing significant challenges. The value of the WUVI decreases consistently throughout 2009. This suggests continuous improvement of the utility's operations.

In Benin, the focus on improved cost recovery resulted in reduced consumption. This resulted in inefficient plant operations, increased losses and costs of operations, and, in some cases, the inability to cope with ongoing urbanization. Benin's poor performance is reflected in the trend of the WUVI in figure 2.7. A value of the WUVI of 80 percent in 2009 suggests increased vulnerability.

In Togo, neglect of cost recovery (illustrated by the downward trend in figure 2.6) increased vulnerability significantly. Togo is hardly balancing its operations, with vulnerability higher than 60 percent throughout the nine-year period depicted in figure 2.7.

Figure 2.6   Cost Recovery for Benin, Gabon, and Togo, 2001–09

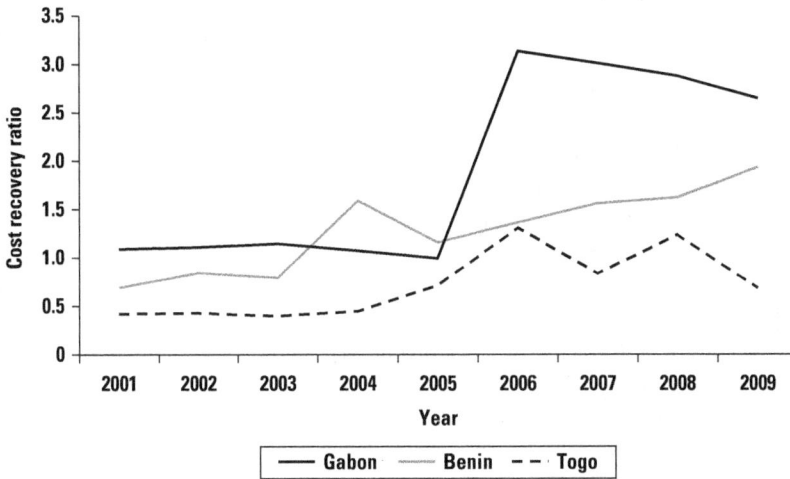

Source: IBNET database.

Figure 2.7   WUVI for Benin, Gabon, and Togo, 2001–09

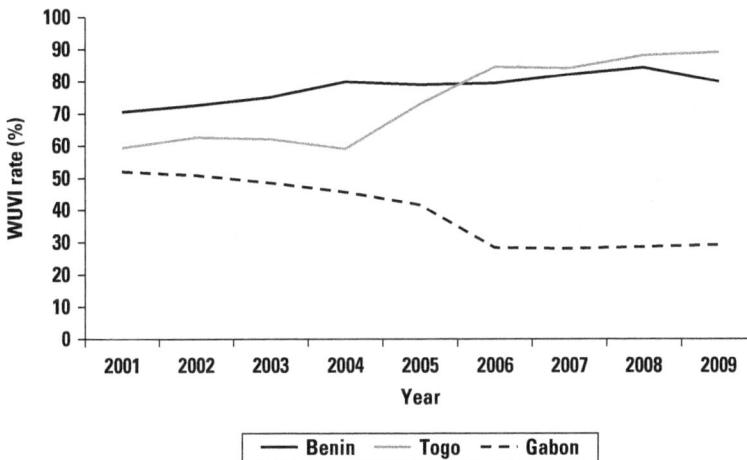

Source: IBNET database.

Czech Republic: Privatization Did Not Affect WUVI. No difference in WUVI values was recorded between private and public utilities in the Czech Republic. All utilities perform very similarly.

Conclusions

1. The IBNET Apgar adequately represents utilities' health, stage of development, and performance status. Its definitions reflect actual statistical distributions that exist in the water sector and are reported to IBNET. Apgar is applicable to utilities at all stages of development and can be used widely for quick assessment of sector performance.

2. WUVI is determined as a dynamic version of the IBNET Apgar. WUVI is an estimated probability that a water utility will experience a performance problem as measured by future Apgar score. WUVI depicts risk, and the higher the threshold that is considered, the more "strict" the index becomes in the sense that the utility must have a high Apgar score to move out of the vulnerability zone. Our conceptualization of a WUVI is as an early warning device rather than an "actionable" index. By this, we mean that a high-value WUVI is a symptom of a possible future problem but does not indicate the specifics of that problem. Hence, we envision that managers and policy makers would treat a high-value WUVI as an indication that further diagnostics are desirable to determine the issues faced by a particular utility and to formulate potential remedies.

3. The combined information provided by the two indices can be used for regulation and rating of the water utilities. Considering the complexity of the sector and the competing demands of its authorities and the different customers, Apgar and WUVI are seen as valuable tools that can assist utility managers and water authorities.

Notes

1. Appearance, Pulse, Grimace, Activity, Respiration. See http://en.wikipedia.org/wiki/Apgar_score.

2. For an example of tracking a uniform product, see the *Economist's* Big Mac Index, which tracks worldwide prices of a Big Mac hamburger: http://bigmacindex.org/2013-big-mac-index.html.

3. The original Apgar score was developed to quickly and summarily assess the health of an infant immediately after childbirth. The Apgar score is determined by evaluating the infant on five simple criteria on a scale of 0 to 2, then summing up the five values thus obtained. Following the original Apgar scores, utilities are then classified as utilities that are "critically low" (with a score of 3.6 or less), "fairly low" (a score between 3.6 and 7.2), or "normal" (a score above 7.2).

References

Cabalu, H. 2010. "Indicators of Security of Natural Gas Supply in Asia." *Energy Policy* 38(1): 218–25.

Council of the European Union. 1998. "Council Directive 98/83/EC of 3 November 1998 on the Quality of Water Intended for Human Consumption. *Official Journal of the European Communities* L 330/32. Available at: http://eur-lex.europa.eu/LexUriServ/LexUriServ.do?uri=CELEX:31998L0083:EN:NOT.

Cutter, S., C. Burton, and C. Emrich. 2010. "Disaster Resilience Indicators for Benchmarking Baseline Conditions." *Journal of Homeland Security and Emergency Management* 7(1): Article 51.

Estrella, A., and F. Mishkin. 1998. "Predicting U.S. Recessions: Financial Variables As Leading Indicators." *Review of Economics and Statistics* 80(1): 45–61.

Gnansounou, E. 2008. "Assessing the Energy Vulnerability: Case of Industrialised Countries." *Energy Policy* 36(10): 3734–44.

Liang, B., and H. Park. 2010. "Predicting Hedge Fund Failure: A Comparison of Risk Measures." *Journal of Financial and Quantitative Analysis* 45(1): 199–222.

Moffitt, L., N. Zirogiannis, and A. Danilenko. 2012. "Developing a Robust Water Utility Vulnerability Index." *Water Asset Management International* 8(1): 6–11.

National Health and Medical Research Council (HMRC), National Resource Management Ministerial Council (NRMMC). 2011. *Australian Drinking Water Guidelines Paper 6 National Water Quality Management Strategy*. National Health and Medical Research Council, National Resource Management Ministerial Council, Commonwealth of Australia, Canberra. Available at: http://www.nhmrc.gov.au/guidelines/publications /eh52.

Standards Organisation of Nigeria (SON). 2007. "Nigerian Standard for Drinking Water Quality." Nigerian Industrial Standard NIS 554: 2007. SON, Lagos, Nigera. Available at: http://www.unicef.org/nigeria/ng_publications_Nigerian_Standard_for_Drinking _Water_Quality.pdf.

van den Berg, C., and A. Danilenko. 2011. *The IBNET Water Supply and Sanitation Performance Blue Book*. Washington, DC: World Bank.

Zirogiannis, N., L. Moffitt, A. Danilenko, R. Rop, and L. Otiego. 2012. "How Healthy Is Your Utility? Consolidating Performance and Vulnerability to Assess Utility Maturity." *Water Utility Management International* 7(2): 19–23.

Appendix. Country Data Tables

IBNET Indicator/Country: Albania

Latest year available	2010	2011	2012
Surface area (km^2)	28,748	28,748	28,748
GNI per capita, Atlas method (current US$)	4,040	4,050	4,030
Total population (thousands)	3,204	3,216	3,228
Urban population (%)	52	53	54
Total urban population (thousands)	1,677	1,717	1,743
MDGs			
Access to improved water sources 2010 (%)[a]	95	95	95
Access to improved sanitation 2010 (%)[a]	94	94	94
IBNET sourced data			
Number of utilities reporting in IBNET sample	58	58	54
Population served (water), (thousands)	2,653	2,678	2,546
Size of the sample: Total population living in service area (water supply), (thousands)	3,308	3,316	3,141
Services coverage			
1.1 Water coverage (%)	80	81	81
2.1 Sewerage coverage (%)	65	65	66
Operational efficiency			
13.2 Electrical energy costs vs. operating costs (%) (share of energy cost as % of operational expenses)	27	26	25
6.1 Nonrevenue water (%)	63.00	64.00	68.00
6.2 Nonrevenue water (m^3/km/day)	76	70	74
12.3 Staff W/1,000 W population served (W/1,000 W population served)	—	—	—
15.1 Continuity of service (hrs/day) (duration of water supply, hours)	11.20	10.80	10.90
Financial efficiency			
8.1 Water sold that is metered (%)	47	58	54
23.1 Collection period (days)	57	28	76
23.2 Collection ratio (%)	152	104	121
18.1 Average revenue W & WW (US$/m^3 water sold)	0.39	0.54	
11.1 Operational cost W & WW (US$/m^3 water sold)	0.59	0.72	0.78
24.1 Operating cost coverage (ratio)	0.67	0.74	0.73
Production and consumption			
3.1 Water production (l/person/day)	304.00	293.00	
4.1 Total water consumption (l/person/day)	111.00	106.00	97.00
4.7 Residential consumption (l/person/day)	72	70	75
Poverty and affordability			
19.1 Total revenues/service population/GNI (% GNI per capita) (average revenues)	0.44	0.55	0.48
19.2 Annual bill for households consuming 6 m^3 of water/month (US$/yr)	10.02	11.00	10.80
21.1 Ratio of industrial to residential tariff (level of cross-subsidy)	2.86	6.19	6.28

a. UNICEF and WHO 2012.

IBNET Indicator/Country: Algeria

Latest year available	2008	2009	2010
Surface area (km²)	2,382,000	2,382,000	2,382,000
GNI per capita, Atlas method (current US$)	4,260	4,170	4,350
Total population (thousands)	34,428	34,950	35,468
Urban population (%)	70	71	72
Total urban population (thousands)	24,062	24,800	25,546
MDGs			
Access to improved water sources 2010 (%)[a]	83	83	83
Access to improved sanitation 2010 (%)[a]	95	95	95
IBNET sourced data			
Number of utilities reporting in IBNET sample	1	1	1
Population served (water), (thousands)	17,445	17,550	19,966
Size of the sample: Total population living in service area (water supply), (thousands)	24,585	25,199	25,829
Services coverage			
1.1 Water coverage (%)	71	70	77
2.1 Sewerage coverage (%)	—	—	—
Operational efficiency			
13.2 Electrical energy costs vs. operating costs (%) (share of energy cost as % of operational expenses)	18	18	18
6.1 Nonrevenue water (%)	55.00	55.00	54.00
6.2 Nonrevenue water (m³/km/day)	58	55	54
12.3 Staff W/1,000 W population served (W/1,000 W population served)	1.20	1.20	1.10
15.1 Continuity of service (hrs/day) (duration of water supply, hours)	—	—	—
Financial efficiency			
8.1 Water sold that is metered (%)	—	—	—
23.1 Collection period (days)	—	—	—
23.2 Collection ratio (%)	—	—	—
18.1 Average revenue W & WW (US$/m³ water sold)	0.39	0.34	0.32
11.1 Operational cost W & WW (US$/m³ water sold)	0.59	0.51	0.49
24.1 Operating cost coverage (ratio)	—	—	—
Production and consumption			
3.1 Water production (l/person/day)	169.00	167.00	152.00
4.1 Total water consumption (l/person/day)	76.00	75.00	69.00
4.7 Residential consumption (l/person/day)	59	58	54
Poverty and affordability			
19.1 Total revenues/service population/GNI (% GNI per capita) (average revenues)	0.45	0.43	0.47
19.2 Annual bill for households consuming 6 m³ of water/month (US$/yr)	—	—	—
21.1 Ratio of industrial to residential tariff (level of cross-subsidy)	1.24	1.20	1.21

a. UNICEF and WHO 2012.

IBNET Indicator/Country: Argentina

Latest year available	2008	2009	2010
Surface area (km^2)	2,780,400	2,780,400	2,780,400
GNI per capita, Atlas method (current US$)	7,190	7,580	8,620
Total population (thousands)	39,714	40,062	40,412
Urban population (%)	92	92	92
Total urban population (thousands)	36,522	36,920	37,320
MDGs			
Access to improved water sources 2010 (%)[a]	98	98	98
Access to improved sanitation 2010 (%)[a]	92	92	92
IBNET sourced data			
Number of utilities reporting in IBNET sample	6	4	5
Population served (water), (thousands)	12,633	11,948	11,522
Size of the sample: Total population living in service area (water supply), (thousands)	14,849	13,375	13,294
Services coverage			
1.1 Water coverage (%)	85	89	87
2.1 Sewerage coverage (%)	64	64	65
Operational efficiency			
13.2 Electrical energy costs vs. operating costs (%) (share of energy cost as % of operational expenses)			
6.1 Nonrevenue water (%)	44.00	40.00	38.00
6.2 Nonrevenue water (m^3/km/day)	99	96	88
12.3 Staff W/1,000 W population served (W/1,000 W population served)			
15.1 Continuity of service (hrs/day) (duration of water supply, hours)	24.00	24.00	24.00
Financial efficiency			
8.1 Water sold that is metered (%)	32	20	20
23.1 Collection period (days)	3	10	2,200
23.2 Collection ratio (%)	100	102	100
18.1 Average revenue W & WW (US$/m^3 water sold)	0.00	0.00	0.00
11.1 Operational cost W & WW (US$/m^3 water sold)	—	—	—
24.1 Operating cost coverage (ratio)	2.20	1.53	0.56
Production and consumption			
3.1 Water production (l/person/day)	558.00	529.00	570.00
4.1 Total water consumption (l/person/day)	324.00	329.00	374.00
4.7 Residential consumption (l/person/day)	129	277	297
Poverty and affordability			
19.1 Total revenues/service population/GNI (% GNI per capita) (average revenues)	0.40	0.45	0.37
19.2 Annual bill for households consuming 6 m^3 of water/month (US$/yr)	—	—	—
21.1 Ratio of industrial to residential tariff (level of cross-subsidy)	1.00	1.00	1.00

a. UNICEF and WHO 2012.

IBNET Indicator/Country: Armenia

Latest year available	2008	2009	2010
Surface area (km²)	29,743	29,743	29,743
GNI per capita, Atlas method (current US$)	3,340	3,440	3,330
Total population (thousands)	3,079	3,085	3,092
Urban population (%)	64	64	64
Total urban population (thousands)	1,974	1,977	1,981
MDGs			
Access to improved water sources 2010 (%)[a]	98	98	98
Access to improved sanitation 2010 (%)[a]	90	90	90
IBNET sourced data			
Number of utilities reporting in IBNET sample	5	5	5
Population served (water), (thousands)	1,752	1,976	2,000
Size of the sample: Total population living in service area (water supply), (thousands)	2,177	2,187	2,188
Services coverage			
1.1 Water coverage (%)	80	90	91
2.1 Sewerage coverage (%)	35	37	37
Operational efficiency			
13.2 Electrical energy costs vs. operating costs (%) (share of energy cost as % of operational expenses)	21	20	16
6.1 Nonrevenue water (%)	84.00	84.00	83.00
6.2 Nonrevenue water (m³/km/day)	95	109	102
12.3 Staff W/1,000 W population served (W/1,000 W population served)	1.60	1.60	1.50
15.1 Continuity of service (hrs/day) (duration of water supply, hours)	13.20	14.00	15.20
Financial efficiency			
8.1 Water sold that is metered (%)	78	83	91
23.1 Collection period (days)	266	296	281
23.2 Collection ratio (%)	87	79	80
18.1 Average revenue W & WW (US$/m³ water sold)	0.47	0.44	0.47
11.1 Operational cost W & WW (US$/m³ water sold)	0.44	0.45	0.47
24.1 Operating cost coverage (ratio)	1.05	0.98	0.98
Production and consumption			
3.1 Water production (l/person/day)	936.00	812.00	759.00
4.1 Total water consumption (l/person/day)	151.00	128.00	126.00
4.7 Residential consumption (l/person/day)	94	85	83
Poverty and affordability			
19.1 Total revenues/service population/GNI (% GNI per capita) (average revenues)	0.78	0.60	0.65
19.2 Annual bill for households consuming 6 m³ of water/month (US$/yr)	32.10	29.74	35.86
21.1 Ratio of industrial to residential tariff (level of cross-subsidy)	1.46	1.45	1.39

a. UNICEF and WHO 2012.

IBNET Indicator/Country: Australia

Latest year available	2009	2010	2011
Surface area (km²)	7,692,024	7,692,024	7,692,024
GNI per capita, Atlas method (current US$)	43,670	46,310	50,150
Total population (thousands)	21,952	22,300	22,621
Urban population (%)	89	89	89
Total urban population (thousands)	19,509	19,857	20,176
MDGs			
Access to improved water sources 2010 (%)[a]	100	100	100
Access to improved sanitation 2010 (%)[a]	100	100	100
IBNET sourced data			
Number of utilities reporting in IBNET sample	60	62	64
Population served (water), (thousands)	15,237	15,820	18,436
Size of the sample: Total population living in service area (water supply), (thousands)	15,224	15,806	18,436
Services coverage			
1.1 Water coverage (%)	100	100	100
2.1 Sewerage coverage (%)	95	94	95
Operational efficiency			
13.2 Electrical energy costs vs. operating costs (%) (share of energy cost as % of operational expenses)	—	—	—
6.1 Nonrevenue water (%)	9.00	10.00	15.00
6.2 Nonrevenue water (m³/km/day)	4	5	6
12.3 Staff W/1,000 W population served (W/1,000 W population served)	—	—	—
15.1 Continuity of service (hrs/day) (duration of water supply, hours)			
Financial efficiency			
8.1 Water sold that is metered (%)	100	100	100
23.1 Collection period (days)	—	—	—
23.2 Collection ratio (%)	—	—	—
18.1 Average revenue W & WW (US$/m³ water sold)	3.26	3.73	5.37
11.1 Operational cost W & WW (US$/m³ water sold)	1.59	1.78	4.07
24.1 Operating cost coverage (ratio)	2.05	2.09	1.32
Production and consumption			
3.1 Water production (l/person/day)	366.00	366.00	315.00
4.1 Total water consumption (l/person/day)	333.00	330.00	285.00
4.7 Residential consumption (l/person/day)	205	207	190
Poverty and affordability			
19.1 Total revenues/service population/GNI (% GNI per capita) (average revenues)	0.91	0.97	1.11
19.2 Annual bill for households consuming 6 m³ of water/month (US$/yr)	—	—	—
21.1 Ratio of industrial to residential tariff (level of cross-subsidy)	—	—	—

a. UNICEF and WHO 2012.

IBNET Indicator/Country: Azerbaijan

Latest year available	2007	2008	2009
Surface area (km^2)	86,600	86,600	86,600
GNI per capita, Atlas method (current US$)	2,710	3,870	4,800
Total population (thousands)	8,581	8,763	8,947
Urban population (%)	53	53	53
Total urban population (thousands)	4,530	4,644	4,760
MDGs			
Access to improved water sources 2010 (%)[a]	80	80	80
Access to improved sanitation 2010 (%)[a]	82	82	82
IBNET sourced data			
Number of utilities reporting in IBNET sample	1	1	1
Population served (water), (thousands)	8,875	8,920	8,973
Size of the sample: Total population living in service area (water supply), (thousands)	8,875	8,920	8,973
Services coverage			
1.1 Water coverage (%)	100	100	100
2.1 Sewerage coverage (%)	56	56	56
Operational efficiency			
13.2 Electrical energy costs vs. operating costs (%) (share of energy cost as % of operational expenses)	—	—	—
6.1 Nonrevenue water (%)	47.00	47.00	47.00
6.2 Nonrevenue water (m^3/km/day)	377	360	355
12.3 Staff W/1,000 W population served (W/1,000 W population served)	0.90	0.90	0.90
15.1 Continuity of service (hrs/day) (duration of water supply, hours)	14.00	14.00	16.00
Financial efficiency			
8.1 Water sold that is metered (%)	9	10	13
23.1 Collection period (days)	107	101	94
23.2 Collection ratio (%)	71	72	74
18.1 Average revenue W & WW (US$/m^3 water sold)	0.36	0.40	0.41
11.1 Operational cost W & WW (US$/m^3 water sold)	0.21	0.24	0.24
24.1 Operating cost coverage (ratio)	1.70	1.69	1.68
Production and consumption			
3.1 Water production (l/person/day)	525.00	525.00	525.00
4.1 Total water consumption (l/person/day)	281.00	281.00	281.00
4.7 Residential consumption (l/person/day)	197	197	197
Poverty and affordability			
19.1 Total revenues/service population/GNI (% GNI per capita) (average revenues)	1.36	1.06	0.88
19.2 Annual bill for households consuming 6 m^3 of water/month (US$/yr)	—	—	—
21.1 Ratio of industrial to residential tariff (level of cross-subsidy)	7.12	7.12	7.12

a. UNICEF and WHO 2012.

IBNET Indicator/Country: Bangladesh

Latest year available	2011	2012	2013
Surface area (km^2)	143,998	143,998	143,998
GNI per capita, Atlas method (current US$)	770	840	880
Total population (thousands)	150,494	152,166	153,837
Urban population (%)	28	29	29
Total urban population (thousands)	42,725	43,885	45,059
MDGs			
Access to improved water sources 2010 (%)[a]	81	81	81
Access to improved sanitation 2010 (%)[a]	56	56	56
IBNET sourced data			
Number of utilities reporting in IBNET sample	33	68	36
Population served (water), (thousands)	15,698	15,866	15,741
Size of the sample: Total population living in service area (water supply), (thousands)	23,528	24,073	25,472
Services coverage			
1.1 Water coverage (%)	63	66	41
2.1 Sewerage coverage (%)	23	16	26
Operational efficiency			
13.2 Electrical energy costs vs. operating costs (%) (share of energy cost as % of operational expenses)	32	39	45
6.1 Nonrevenue water (%)	32.00	29.00	26.00
6.2 Nonrevenue water (m^3/km/day)	116	101	94
12.3 Staff W/1,000 W population served (W/1,000 W population served)	0.30	0.30	0.30
15.1 Continuity of service (hrs/day) (duration of water supply, hours)	12.00	12.00	8.50
Financial efficiency			
8.1 Water sold that is metered (%)	70	74	83
23.1 Collection period (days)	205	189	183
23.2 Collection ratio (%)	84	81	85
18.1 Average revenue W & WW (US$/m^3 water sold)	0.14	0.15	0.16
11.1 Operational cost W & WW (US$/m^3 water sold)	0.11	0.10	0.11
24.1 Operating cost coverage (ratio)	0.24	0.54	1.44
Production and consumption			
3.1 Water production (l/person/day)	94.00	87.00	161.00
4.1 Total water consumption (l/person/day)	108.00	113.00	119.00
4.7 Residential consumption (l/person/day)	96	101	106
Poverty and affordability			
19.1 Total revenues/service population/GNI (% GNI per capita) (average revenues)	0.72	0.74	0.79
19.2 Annual bill for households consuming 6 m^3 of water/month (US$/yr)	0.33	0.15	11.78
21.1 Ratio of industrial to residential tariff (level of cross-subsidy)	1.10	1.20	3.02

a. UNICEF and WHO 2012.

IBNET Indicator/Country: Bahrain

Latest year available	2008	2009	2010
Surface area (km^2)	665	665	665
GNI per capita, Atlas method (current US$)	18,730	15,590	14,820
Total population (thousands)	1,052	1,170	1,262
Urban population (%)	89	89	89
Total urban population (thousands)	932	1,036	1,118
MDGs			
Access to improved water sources 2010 (%)[a]	100	100	100
Access to improved sanitation 2010 (%)[a]	89	89	89
IBNET sourced data			
Number of utilities reporting in IBNET sample	1	1	1
Population served (water), (thousands)	1,122	1,159	1,235
Size of the sample: Total population living in service area (water supply), (thousands)	1,122	1,159	1,235
Services coverage			
1.1 Water coverage (%)	100	100	100
2.1 Sewerage coverage (%)	—	—	—
Operational efficiency			
13.2 Electrical energy costs vs. operating costs (%) (share of energy cost as % of operational expenses)	—	—	—
6.1 Nonrevenue water (%)	44.00	46.00	40.00
6.2 Nonrevenue water (m^3/km/day)	61	67	59
12.3 Staff W/1,000 W population served (W/1,000 W population served)	0.90	0.90	0.80
15.1 Continuity of service (hrs/day) (duration of water supply, hours)	24.00	24.00	24.00
Financial efficiency			
8.1 Water sold that is metered (%)	100	100	100
23.1 Collection period (days)	—	—	—
23.2 Collection ratio (%)	—	—	—
18.1 Average revenue W & WW (US$/m^3 water sold)	0.26	0.24	0.29
11.1 Operational cost W & WW (US$/m^3 water sold)	1.72	1.94	1.62
24.1 Operating cost coverage (ratio)	0.15	0.13	0.18
Production and consumption			
3.1 Water production (l/person/day)	491.00	552.00	526.00
4.1 Total water consumption (l/person/day)	275.00	280.00	315.00
4.7 Residential consumption (l/person/day)	231	238	257
Poverty and affordability			
19.1 Total revenues/service population/GNI (% GNI per capita) (average revenues)	0.14	0.16	0.22
19.2 Annual bill for households consuming 6 m^3 of water/month (US$/yr)	4.86	4.78	4.80
21.1 Ratio of industrial to residential tariff (level of cross-subsidy)	4.16	4.69	6.57

a. UNICEF and WHO 2012.

IBNET Indicator/Country: Belarus

Latest year available	2010	2011	2012
Surface area (km²)	207,600	207,600	207,600
GNI per capita, Atlas method (current US$)	5,990	6,270	6,530
Total population (thousands)	9,490	9,473	9,455
Urban population (%)	75	75	75
Total urban population (thousands)	7,081	7,107	7,134
MDGs			
Access to improved water sources 2010 (%)[a]	100	100	100
Access to improved sanitation 2010 (%)[a]	93	93	93
IBNET sourced data			
Number of utilities reporting in IBNET sample	20	17	21
Population served (water), (thousands)	1,562	1,904	2,014
Size of the sample: Total population living in service area (water supply), (thousands)	1,630	2,006	2,152
Services coverage			
1.1 Water coverage (%)	91	95	94
2.1 Sewerage coverage (%)	78	75	78
Operational efficiency			
13.2 Electrical energy costs vs. operating costs (%) (share of energy cost as % of operational expenses)	27	31	28
6.1 Nonrevenue water (%)	15.00	26	27
6.2 Nonrevenue water (m³/km/day)	16	21.40	20.10
12.3 Staff W/1,000 W population served (W/1,000 W population served)	1.50	1.30	
15.1 Continuity of service (hrs/day) (duration of water supply, hours)	24.00	21.60	24.00
Financial efficiency			
8.1 Water sold that is metered (%)	81	94	94
23.1 Collection period (days)	52	54	55
23.2 Collection ratio (%)	84	100	81
18.1 Average revenue W & WW (US$/m³ water sold)	0.76	1.01	1.05
11.1 Operational cost W & WW (US$/m³ water sold)	0.62	0.85	0.86
24.1 Operating cost coverage (ratio)	1.23	1.19	1.22
Production and consumption			
3.1 Water production (l/person/day)	235.00	234.00	230.00
4.1 Total water consumption (l/person/day)	241.00	174.00	168.00
4.7 Residential consumption (l/person/day)	183	121	117
Poverty and affordability			
19.1 Total revenues/service population/GNI (% GNI per capita) (average revenues)	1.12	1.02	0.99
19.2 Annual bill for households consuming 6 m³ of water/month (US$/yr)	13.51	13.61	10.45
21.1 Ratio of industrial to residential tariff (level of cross-subsidy)	13.08	8.27	7.58

a. UNICEF and WHO 2012.

IBNET Indicator/Country: Benin

Latest year available	2007	2008	2009
Surface area (km^2)	112,622	112,622	112,622
GNI per capita, Atlas method (current US$)	630	730	710
Total population (thousands)	8,113	8,356	8,602
Urban population (%)	42	43	44
Total urban population (thousands)	3,432	3,589	3,751
MDGs			
Access to improved water sources 2010 (%)[a]	75	75	75
Access to improved sanitation 2010 (%)[a]	13	13	13
IBNET sourced data			
Number of utilities reporting in IBNET sample	1	1	1
Population served (water), (thousands)	1,598	1,703	1,860
Size of the sample: Total population living in service area (water supply), (thousands)	3,070	3,170	3,270
Services coverage			
1.1 Water coverage (%)	52	54	57
2.1 Sewerage coverage (%)	—	—	—
Operational efficiency			
13.2 Electrical energy costs vs. operating costs (%) (share of energy cost as % of operational expenses)	23	15	20
6.1 Nonrevenue water (%)	28.00	24.00	28.00
6.2 Nonrevenue water (m^3/km/day)	6	5	6
12.3 Staff W/1,000 W population served (W/1,000 W population served)	0.50	0.40	0.40
15.1 Continuity of service (hrs/day) (duration of water supply, hours)	—	—	—
Financial efficiency			
8.1 Water sold that is metered (%)	100	100	100
23.1 Collection period (days)	190	219	199
23.2 Collection ratio (%)	102	93	91
18.1 Average revenue W & WW (US$/m^3 water sold)	1.17	1.28	1.37
11.1 Operational cost W & WW (US$/m^3 water sold)	0.74	0.78	0.70
24.1 Operating cost coverage (ratio)	1.58	1.64	1.96
Production and consumption			
3.1 Water production (l/person/day)	58.00	59.00	57.00
4.1 Total water consumption (l/person/day)	42.00	45.00	41.00
4.7 Residential consumption (l/person/day)	—	—	—
Poverty and affordability			
19.1 Total revenues/service population/GNI (% GNI per capita) (average revenues)	2.85	2.88	2.89
19.2 Annual bill for households consuming 6 m^3 of water/month (US$/yr)	—	—	—
21.1 Ratio of industrial to residential tariff (level of cross-subsidy)	—	—	—

a. UNICEF and WHO 2012.

IBNET Indicator/Country: Bhutan

Latest year available	2002	2003	2004
Surface area (km^2)	38,394	38,394	38,394
GNI per capita, Atlas method (current US$)	850	940	1,060
Total population (thousands)	606	624	642
Urban population (%)	28	29	30
Total urban population (thousands)	168	180	192
MDGs			
Access to improved water sources 2010 (%)[a]	96	96	96
Access to improved sanitation 2010 (%)[a]	44	44	44
IBNET sourced data			
Number of utilities reporting in IBNET sample	1	1	1
Population served (water), (thousands)	40.00	42.00	43.00
Size of the sample: Total population living in service area (water supply), (thousands)	60	60	60
Services coverage			
1.1 Water coverage (%)	67	70	72
2.1 Sewerage coverage (%)	0	0	0
Operational efficiency			
13.2 Electrical energy costs vs. operating costs (%) (share of energy cost as % of operational expenses)	—	—	—
6.1 Nonrevenue water (%)	38.00	47.00	46.00
6.2 Nonrevenue water (m^3/km/day)	48	69	68
12.3 Staff W/1,000 W population served (W/1,000 W population served)	1.40	1.30	1.30
15.1 Continuity of service (hrs/day) (duration of water supply, hours)	13.00	13.00	13.00
Financial efficiency			
8.1 Water sold that is metered (%)	—	—	—
23.1 Collection period (days)	—	—	—
23.2 Collection ratio (%)	81	73	88
18.1 Average revenue W & WW (US$/m^3 water sold)	0.04	0.06	0.06
11.1 Operational cost W & WW (US$/m^3 water sold)	0.03	0.04	0.04
24.1 Operating cost coverage (ratio)	1.28	1.55	1.55
Production and consumption			
3.1 Water production (l/person/day)	251.61	283.02	279.63
4.1 Total water consumption (l/person/day)	156.00	150.00	151.00
4.7 Residential consumption (l/person/day)	105	101	102
Poverty and affordability			
19.1 Total revenues/service population/GNI (% GNI per capita) (average revenues)	0.27	0.35	0.31
19.2 Annual bill for households consuming 6 m^3 of water/month (US$/yr)	—	—	—
21.1 Ratio of industrial to residential tariff (level of cross-subsidy)	0.95	0.83	0.82

a. UNICEF and WHO 2012.

IBNET Indicator/Country: Plurinational State of Bolivia

Latest year available	2004	2005	2006
Surface area (km²)	1,098,581	1,098,581	1,098,581
GNI per capita, Atlas method (current US$)	960	1,030	1,120
Total population (thousands)	8,983	9,147	9,307
Urban population (%)	64	64	65
Total urban population (thousands)	5,724	5,872	6,018
MDGs			
Access to improved water sources 2010 (%)[a]	88	88	88
Access to improved sanitation 2010 (%)[a]	27	27	27
IBNET sourced data			
Number of utilities reporting in IBNET sample	2	2	5
Population served (water), (thousands)	2,321.00	2,388.00	2,155.00
Size of the sample: Total population living in service area (water supply), (thousands)	2,355	2,510	2,453
Services coverage			
1.1 Water coverage (%)	99	95	88
2.1 Sewerage coverage (%)	69	64	66
Operational efficiency			
13.2 Electrical energy costs vs. operating costs (%) (share of energy cost as % of operational expenses)	—	6	23
6.1 Nonrevenue water (%)	28.00	28.00	35.00
6.2 Nonrevenue water (m³/km/day)	13	17	24
12.3 Staff W/1,000 W population served (W/1,000 W population served)	—	0.20	0.80
15.1 Continuity of service (hrs/day) (duration of water supply, hours)	24.00	24.00	20.00
Financial efficiency			
8.1 Water sold that is metered (%)	—	100	92
23.1 Collection period (days)	—	117	72
23.2 Collection ratio (%)	—	91	723
18.1 Average revenue W & WW (US$/m³ water sold)	0.67	0.45	0.40
11.1 Operational cost W & WW (US$/m³ water sold)	0.58	0.44	0.26
24.1 Operating cost coverage (ratio)	1.31	1.02	1.56
Production and consumption			
3.1 Water production (l/person/day)	100.00	129.17	127.69
4.1 Total water consumption (l/person/day)	72.00	93.00	83.00
4.7 Residential consumption (l/person/day)	94	78	61
Poverty and affordability			
19.1 Total revenues/service population/GNI (% GNI per capita) (average revenues)	1.83	1.48	1.08
19.2 Annual bill for households consuming 6 m³ of water/month (US$/yr)	—	—	25.88
21.1 Ratio of industrial to residential tariff (level of cross-subsidy)	—	—	3.26

a. UNICEF and WHO 2012.

IBNET Indicator/Country: Bosnia and Herzegovina

Latest year available	2005	2006	2007
Surface area (km^2)	51,209	51,209	51,209
GNI per capita, Atlas method (current US$)	3,020	3,350	3,820
Total population (thousands)	3,781	3,782	3,779
Urban population (%)	45	46	46
Total urban population (thousands)	1,711	1,730	1,748
MDGs			
Access to improved water sources 2010 (%)[a]	99	99	99
Access to improved sanitation 2010 (%)[a]	95	95	95
IBNET sourced data			
Number of utilities reporting in IBNET sample	20	20	20
Population served (water), (thousands)	1,179	1,185	1,230
Size of the sample: Total population living in service area (water supply), (thousands)	1,301	1,279	1,328
Services coverage			
1.1 Water coverage (%)	91	93	93
2.1 Sewerage coverage (%)	56	56	55
Operational efficiency			
13.2 Electrical energy costs vs. operating costs (%) (share of energy cost as % of operational expenses)	13	13	119
6.1 Nonrevenue water (%)	61.00	62.00	60.00
6.2 Nonrevenue water (m^3/km/day)	77	63	60
12.3 Staff W/1,000 W population served (W/1,000 W population served)	1.40	1.30	1.30
15.1 Continuity of service (hrs/day) (duration of water supply, hours)	23.20	23.30	24.00
Financial efficiency			
8.1 Water sold that is metered (%)	99	98	99
23.1 Collection period (days)	227	246	334
23.2 Collection ratio (%)	79	83	159
18.1 Average revenue W & WW (US$/m^3 water sold)	0.71	0.77	0.82
11.1 Operational cost W & WW (US$/m^3 water sold)	0.60	0.80	0.84
24.1 Operating cost coverage (ratio)	1.05	0.94	0.97
Production and consumption			
3.1 Water production (l/person/day)	464.00	423.00	411.00
4.1 Total water consumption (l/person/day)	183.00	162.00	164.00
4.7 Residential consumption (l/person/day)	137	121	122
Poverty and affordability			
19.1 Total revenues/service population/GNI (% GNI per capita) (average revenues)	1.57	1.36	1.28
19.2 Annual bill for households consuming 6 m^3 of water/month (US$/yr)	44.59	46.39	53.72
21.1 Ratio of industrial to residential tariff (level of cross-subsidy)	2.79	2.94	2.73

a. UNICEF and WHO 2012.

IBNET Indicator/Country: Brazil

Latest year available	2009	2010	2011
Surface area (km²)	8,514,877	8,514,877	8,514,877
GNI per capita, Atlas method (current US$)	8,140	9,520	10,700
Total population (thousands)	193,247	194,946	196,655
Urban population (%)	84	84	85
Total urban population (thousands)	162,394	164,408	166,370
MDGs			
Access to improved water sources 2010 (%)[a]	98	98	98
Access to improved sanitation 2010 (%)[a]	79	79	79
IBNET sourced data			
Number of utilities reporting in IBNET sample	920	985	1,063
Population served (water), (thousands)	111,875	150,983	190,091
Size of the sample: Total population living in service area (water supply), (thousands)	144,785	147,709	150,633
Services coverage			
1.1 Water coverage (%)	80	81	81
2.1 Sewerage coverage (%)	42	43	47
Operational efficiency			
13.2 Electrical energy costs vs. operating costs (%) (share of energy cost as % of operational expenses)	—	—	—
6.1 Nonrevenue water (%)	40.00	39.00	39.00
6.2 Nonrevenue water (m³/km/day)	35	33	33
12.3 Staff W/1,000 W population served (W/1,000 W population served)	—	—	
15.1 Continuity of service (hrs/day) (duration of water supply, hours)	24.00	24.00	24.00
Financial efficiency			
8.1 Water sold that is metered (%)	94	—	—
23.1 Collection period (days)	115	111	138
23.2 Collection ratio (%)	92	93	99
18.1 Average revenue W & WW (US$/m³ water sold)	1.49	1.67	2.03
11.1 Operational cost W & WW (US$/m³ water sold)	1.38	1.10	1.41
24.1 Operating cost coverage (ratio)	1.08	1.52	1.44
Production and consumption			
3.1 Water production (l/person/day)	274.00	284.00	
4.1 Total water consumption (l/person/day)	169.00	168.00	174.00
4.7 Residential consumption (l/person/day)	—	110	116
Poverty and affordability			
19.1 Total revenues/service population/GNI (% GNI per capita) (average revenues)	1.13	1.08	1.20
19.2 Annual bill for households consuming 6 m³ of water/month (US$/yr)	—	—	—
21.1 Ratio of industrial to residential tariff (level of cross-subsidy)	1.00	1.00	1.00

a. UNICEF and WHO 2012.

IBNET Indicator/Country: Bulgaria

Latest year available	2006	2007	2008
Surface area (km^2)	110,879	110,879	110,879
GNI per capita, Atlas method (current US$)	4,080	4,530	5,700
Total population (thousands)	7,699	7,660	7,623
Urban population (%)	71	71	71
Total urban population (thousands)	5,428	5,423	5,420
MDGs			
Access to improved water sources 2010 (%)[a]	100	100	100
Access to improved sanitation 2010 (%)[a]	100	100	100
IBNET sourced data			
Number of utilities reporting in IBNET sample	20	20	20
Population served (water), (thousands)	5,246	5,398	5,389
Size of the sample: Total population living in service area (water supply), (thousands)	5,288	5,436	5,422
Services coverage			
1.1 Water coverage (%)	99	99	99
2.1 Sewerage coverage (%)	60	61	61
Operational efficiency			
13.2 Electrical energy costs vs. operating costs (%) (share of energy cost as % of operational expenses)	12	11	11
6.1 Nonrevenue water (%)	58.00	56.00	54.00
6.2 Nonrevenue water (m^3/km/day)	31	27	26
12.3 Staff W/1,000 W population served (W/1,000 W population served)	1.60	1.60	1.60
15.1 Continuity of service (hrs/day) (duration of water supply, hours)	24.00	24.00	24.00
Financial efficiency			
8.1 Water sold that is metered (%)	97	98	99
23.1 Collection period (days)	141	130	98
23.2 Collection ratio (%)	136	128	124
18.1 Average revenue W & WW (US$/m^3 water sold)	0.68	0.78	1.01
11.1 Operational cost W & WW (US$/m^3 water sold)	0.50	0.58	0.77
24.1 Operating cost coverage (ratio)	1.39	1.35	1.32
Production and consumption			
3.1 Water production (l/person/day)	416.00	385.00	372.00
4.1 Total water consumption (l/person/day)	174.00	171.00	171.00
4.7 Residential consumption (l/person/day)	159	152	151
Poverty and affordability			
19.1 Total revenues/service population/GNI (% GNI per capita) (average revenues)	1.06	1.07	1.11
19.2 Annual bill for households consuming 6 m^3 of water/month (US$/yr)	44.65	55.92	72.33
21.1 Ratio of industrial to residential tariff (level of cross-subsidy)	1.19	1.24	1.10

a. UNICEF and WHO 2012.

IBNET Indicator/Country: Burkina Faso

Latest year available	2007	2008	2009
Surface area (km^2)	274,222	274,222	274,222
GNI per capita, Atlas method (current US$)	430	480	550
Total population (thousands)	15,061	15,515	15,984
Urban population (%)	23	24	25
Total urban population (thousands)	3,493	3,726	3,971
MDGs			
Access to improved water sources 2010 (%)[a]	79	79	79
Access to improved sanitation 2010 (%)[a]	17	17	17
IBNET sourced data			
Number of utilities reporting in IBNET sample	1	1	1
Population served (water), (thousands)	2,330	2,508	2,518
Size of the sample: Total population living in service area (water supply), (thousands)	3,178	3,351	3,509
Services coverage			
1.1 Water coverage (%)	73	75	72
2.1 Sewerage coverage (%)	0	0	0
Operational efficiency			
13.2 Electrical energy costs vs. operating costs (%) (share of energy cost as % of operational expenses)	21	20	20
6.1 Nonrevenue water (%)	18.00	18.00	18.00
6.2 Nonrevenue water (m^3/km/day)	5	5	5
12.3 Staff W/1,000 W population served (W/1,000 W population served)	0.30	0.20	0.20
15.1 Continuity of service (hrs/day) (duration of water supply, hours)	23.00	23.00	23.00
Financial efficiency			
8.1 Water sold that is metered (%)	100	100	100
23.1 Collection period (days)	—	—	—
23.2 Collection ratio (%)	—	—	—
18.1 Average revenue W & WW (US$/m^3 water sold)	1.57	1.89	1.67
11.1 Operational cost W & WW (US$/m^3 water sold)	0.73	0.86	0.81
24.1 Operating cost coverage (ratio)	2.15	2.19	2.07
Production and consumption			
3.1 Water production (l/person/day)	58.00	58.00	61.00
4.1 Total water consumption (l/person/day)	47.00	47.00	50.00
4.7 Residential consumption (l/person/day)	40	40	42
Poverty and affordability			
19.1 Total revenues/service population/GNI (% GNI per capita) (average revenues)	6.26	6.75	5.54
19.2 Annual bill for households consuming 6 m^3 of water/month (US$/yr)	—	—	—
21.1 Ratio of industrial to residential tariff (level of cross-subsidy)	3.49	3.38	3.36

a. UNICEF and WHO 2012.

IBNET Indicator/Country: Burundi

Latest year available	2004	2005	2006
Surface area (km^2)	27,834	27,834	27,834
GNI per capita, Atlas method (current US$)	130	140	160
Total population (thousands)	7,040	7,251	7,474
Urban population (%)	9	9	10
Total urban population (thousands)	652	653	732
MDGs			
Access to improved water sources 2010 (%)[a]	72	72	72
Access to improved sanitation 2010 (%)[a]	46	46	46
IBNET sourced data			
Number of utilities reporting in IBNET sample	1	1	1
Population served (water), (thousands)	650	700	750
Size of the sample: Total population living in service area (water supply), (thousands)	6,000	6,500	7,000
Services coverage			
1.1 Water coverage (%)	11	11	11
2.1 Sewerage coverage (%)	—	—	—
Operational efficiency			
13.2 Electrical energy costs vs. operating costs (%) (share of energy cost as % of operational expenses)	—	—	—
6.1 Nonrevenue water (%)	45.00	40.00	40.00
6.2 Nonrevenue water (m^3/km/day)	20	17	17
12.3 Staff W/1,000 W population served (W/1,000 W population served)	0.60	0.60	0.70
15.1 Continuity of service (hrs/day) (duration of water supply, hours)	15.00	15.00	15.00
Financial efficiency			
8.1 Water sold that is metered (%)	—	—	—
23.1 Collection period (days)	430	330	250
23.2 Collection ratio (%)	97	100	97
18.1 Average revenue W & WW (US$/m^3 water sold)	0.21	0.21	0.24
11.1 Operational cost W & WW (US$/m^3 water sold)	0.09	0.08	0.09
24.1 Operating cost coverage (ratio)	2.49	2.60	2.76
Production and consumption			
3.1 Water production (l/person/day)	141.82	128.33	121.67
4.1 Total water consumption (l/person/day)	78.00	77.00	73.00
4.7 Residential consumption (l/person/day)	38	36	35
Poverty and affordability			
19.1 Total revenues/service population/GNI (% GNI per capita) (average revenues)	4.60	4.22	4.00
19.2 Annual bill for households consuming 6 m^3 of water/month (US$/yr)	—	—	—
21.1 Ratio of industrial to residential tariff (level of cross-subsidy)	1.72	1.59	1.66

a. UNICEF and WHO 2012.

IBNET Indicator/Country: Cabo Verde

Latest year available	2003	2004	2005
Surface area (km^2)	4,033	4,033	4,033
GNI per capita, Atlas method (current US$)	1,490	1,800	2,080
Total population (thousands)	460	467	473
Urban population (%)	56	57	58
Total urban population (thousands)	258	265	273
MDGs			
Access to improved water sources 2010 (%)[a]	88	88	88
Access to improved sanitation 2010 (%)[a]	61	61	61
IBNET sourced data			
Number of utilities reporting in IBNET sample	1	1	1
Population served (water), (thousands)	91	101	107
Size of the sample: Total population living in service area (water supply), (thousands)	215	223	232
Services coverage			
1.1 Water coverage (%)	42	45	46
2.1 Sewerage coverage (%)	—	—	—
Operational efficiency			
13.2 Electrical energy costs vs. operating costs (%) (share of energy cost as % of operational expenses)	—	—	—
6.1 Nonrevenue water (%)	30.00	30.00	31.00
6.2 Nonrevenue water (m^3/km/day)	10	11	11
12.3 Staff W/1,000 W population served (W/1,000 W population served)	7.20	6.20	5.90
15.1 Continuity of service (hrs/day) (duration of water supply, hours)	—	—	—
Financial efficiency			
8.1 Water sold that is metered (%)	—	—	—
23.1 Collection period (days)	—	—	—
23.2 Collection ratio (%)	—	—	—
18.1 Average revenue W & WW (US$/m^3 water sold)	3.07	3.52	3.49
11.1 Operational cost W & WW (US$/m^3 water sold)	—	—	—
24.1 Operating cost coverage (ratio)	—	—	—
Production and consumption			
3.1 Water production (l/person/day)	122.86	110.00	108.70
4.1 Total water consumption (l/person/day)	86.00	77.00	75.00
4.7 Residential consumption (l/person/day)	50	49	49
Poverty and affordability			
19.1 Total revenues/service population/GNI (% GNI per capita) (average revenues)	6.47	5.50	4.59
19.2 Annual bill for households consuming 6 m^3 of water/month (US$/yr)	—	—	—
21.1 Ratio of industrial to residential tariff (level of cross-subsidy)	1.00	1.00	1.00

a. UNICEF and WHO 2012.

IBNET Indicator/Country: Cambodia

Latest year available	2005	2006	2007
Surface area (km^2)	181,035	181,035	181,035
GNI per capita, Atlas method (current US$)	460	520	590
Total population (thousands)	13,358	13,516	13,670
Urban population (%)	19	19	19
Total urban population (thousands)	2,538	2,568	2,597
MDGs			
Access to improved water sources 2010 (%)[a]	64	64	64
Access to improved sanitation 2010 (%)[a]	31	31	31
IBNET sourced data			
Number of utilities reporting in IBNET sample	1	1	1
Population served (water), (thousands)	830	910	1,068
Size of the sample: Total population living in service area (water supply), (thousands)	1,106	1,214	1,335
Services coverage			
1.1 Water coverage (%)	75	75	80
2.1 Sewerage coverage (%)	—	—	—
Operational efficiency			
13.2 Electrical energy costs vs. operating costs (%) (share of energy cost as % of operational expenses)	35	45	47
6.1 Nonrevenue water (%)	9.00	7.00	6.00
6.2 Nonrevenue water (m^3/km/day)	12	10	8
12.3 Staff W/1,000 W population served (W/1,000 W population served)	0.60	0.60	0.50
15.1 Continuity of service (hrs/day) (duration of water supply, hours)	24.00	24.00	24.00
Financial efficiency			
8.1 Water sold that is metered (%)	100	100	100
23.1 Collection period (days)	89	94	67
23.2 Collection ratio (%)	—	—	—
18.1 Average revenue W & WW (US$/m^3 water sold)	0.24	0.20	0.28
11.1 Operational cost W & WW (US$/m^3 water sold)	0.11	0.10	0.12
24.1 Operating cost coverage (ratio)	2.24	2.08	2.36
Production and consumption			
3.1 Water production (l/person/day)	204.40	211.83	182.98
4.1 Total water consumption (l/person/day)	186.00	197.00	172.00
4.7 Residential consumption (l/person/day)	113	118	101
Poverty and affordability			
19.1 Total revenues/service population/GNI (% GNI per capita) (average revenues)	3.54	2.77	2.98
19.2 Annual bill for households consuming 6 m^3 of water/month (US$/yr)	9.68	9.41	9.69
21.1 Ratio of industrial to residential tariff (level of cross-subsidy)	1.34	1.36	1.32

a. UNICEF and WHO 2012.

IBNET Indicator/Country: Cameroon

Latest year available	2008	2009
Surface area (km^2)	475,442	475,442
GNI per capita, Atlas method (current US$)	1,150	1,150
Total population (thousands)	18,759	19,175
Urban population (%)	50	51
Total urban population (thousands)	9,440	9,764
MDGs		
Access to improved water sources 2010 (%)[a]	77	77
Access to improved sanitation 2010 (%)[a]	49	49
IBNET sourced data		
Number of utilities reporting in IBNET sample	1	1
Population served (water), (thousands)	3,261	3,500
Size of the sample: Total population living in service area (water supply), (thousands)	8,000	8,300
Services coverage		
1.1 Water coverage (%)	41	42
2.1 Sewerage coverage (%)	—	—
Operational efficiency		
13.2 Electrical energy costs vs. operating costs (%) (share of energy cost as % of operational expenses)	23	23
6.1 Nonrevenue water (%)	45.00	40.00
6.2 Nonrevenue water (m^3/km/day)	327	426
12.3 Staff W/1,000 W population served (W/1,000 W population served)	0.70	0.60
15.1 Continuity of service (hrs/day) (duration of water supply, hours)	24.00	24.00
Financial efficiency		
8.1 Water sold that is metered (%)	100	100
23.1 Collection period (days)	—	—
23.2 Collection ratio (%)	113	117
18.1 Average revenue W & WW (US$/m^3 water sold)	0.76	0.88
11.1 Operational cost W & WW (US$/m^3 water sold)	0.57	0.73
24.1 Operating cost coverage (ratio)	1.34	1.20
Production and consumption		
3.1 Water production (l/person/day)	70.00	97.00
4.1 Total water consumption (l/person/day)	39.00	59.00
4.7 Residential consumption (l/person/day)	24	37
Poverty and affordability		
19.1 Total revenues/service population/GNI (% GNI per capita) (average revenues)	0.94	1.65
19.2 Annual bill for households consuming 6 m^3 of water/month (US$/yr)	—	—
21.1 Ratio of industrial to residential tariff (level of cross-subsidy)	1.00	1.00

a. UNICEF and WHO 2012.

IBNET Indicator/Country: Central African Republic

Latest year available	2007	2008	2009
Surface area (km^2)	622,984	622,984	622,984
GNI per capita, Atlas method (current US$)	380	420	480
Total population (thousands)	4,161	4,238	4,318
Urban population (%)	38	39	39
Total urban population (thousands)	1,597	1,633	1,671
MDGs			
Access to improved water sources 2010 (%)[a]	67	67	67
Access to improved sanitation 2010 (%)[a]	34	34	34
IBNET sourced data			
Number of utilities reporting in IBNET sample	1	1	1
Population served (water), (thousands)	876	897	920
Size of the sample: Total population living in service area (water supply), (thousands)	310	315	324
Services coverage			
1.1 Water coverage (%)	35	35	35
2.1 Sewerage coverage (%)	—	—	—
Operational efficiency			
13.2 Electrical energy costs vs. operating costs (%) (share of energy cost as % of operational expenses)			
6.1 Nonrevenue water (%)	53.00	51.00	51.00
6.2 Nonrevenue water (m^3/km/day)	46	44	—
12.3 Staff W/1,000 W population served (W/1,000 W population served)	0.70	0.70	0.70
15.1 Continuity of service (hrs/day) (duration of water supply, hours)			
Financial efficiency			
8.1 Water sold that is metered (%)	89	77	78
23.1 Collection period (days)	—	—	—
23.2 Collection ratio (%)	67	83	86
18.1 Average revenue W & WW (US$/m^3 water sold)	0.76	0.82	0.71
11.1 Operational cost W & WW (US$/m^3 water sold)	0.45	0.58	—
24.1 Operating cost coverage (ratio)	0.67	1.42	—
Production and consumption			
3.1 Water production (l/person/day)			
4.1 Total water consumption (l/person/day)	42.00	42.00	38.00
4.7 Residential consumption (l/person/day)	19	16	16
Poverty and affordability			
19.1 Total revenues/service population/GNI (% GNI per capita) (average revenues)	3.07	2.99	2.05
19.2 Annual bill for households consuming 6 m^3 of water/month (US$/yr)	—	—	—
21.1 Ratio of industrial to residential tariff (level of cross-subsidy)	—	1.00	1.00

a. UNICEF and WHO 2012.

IBNET Indicator/Country: Chile

Latest year available	2006	2007	2008
Surface area (km²)	756,102	756,102	756,102
GNI per capita, Atlas method (current US$)	7,280	8,620	2,480
Total population (thousands)	16,469	16,633	16,796
Urban population (%)	88	88	88
Total urban population (thousands)	14,492	14,571	14,760
MDGs			
Access to improved water sources 2010 (%)[a]	96	96	96
Access to improved sanitation 2010 (%)[a]	96	96	96
IBNET sourced data			
Number of utilities reporting in IBNET sample	18	18	18
Population served (water), (thousands)	13,310	13,733	14,118
Size of the sample: Total population living in service area (water supply), (thousands)	13,340	13,756	14,165
Services coverage			
1.1 Water coverage (%)	100	100	100
2.1 Sewerage coverage (%)	95	95	95
Operational efficiency			
13.2 Electrical energy costs vs. operating costs (%) (share of energy cost as % of operational expenses)	—	—	—
6.1 Nonrevenue water (%)	34.00	35.00	33.00
6.2 Nonrevenue water (m³/km/day)	37	39	37
12.3 Staff W/1,000 W population served (W/1,000 W population served)	—	—	—
15.1 Continuity of service (hrs/day) (duration of water supply, hours)	24.00	24.00	24.00
Financial efficiency			
8.1 Water sold that is metered (%)	98	98	98
23.1 Collection period (days)	73	76	78
23.2 Collection ratio (%)	107	100	100
18.1 Average revenue W & WW (US$/m³ water sold)	1.01	1.14	1.25
11.1 Operational cost W & WW (US$/m³ water sold)	0.61	0.56	0.56
24.1 Operating cost coverage (ratio)	1.66	2.03	2.24
Production and consumption			
3.1 Water production (l/person/day)	297.00	292.00	285.00
4.1 Total water consumption (l/person/day)	198.00	190.00	190.00
4.7 Residential consumption (l/person/day)	150	144	140
Poverty and affordability			
19.1 Total revenues/service population/GNI (% GNI per capita) (average revenues)	1.00	0.92	3.50
19.2 Annual bill for households consuming 6 m³ of water/month (US$/yr)	—	—	—
21.1 Ratio of industrial to residential tariff (level of cross-subsidy)	1.22	1.30	1.20

a. UNICEF and WHO 2012.

IBNET Indicator/Country: China

Latest year available	2010	2011	2012
Surface area (km^2)	9,640,821	9,640,821	9,640,821
GNI per capita, Atlas method (current US$)	4,240	4,900	5,720
Total population (thousands)	1,337,705	1,344,120	1,350,622
Urban population (%)	49	51	52
Total urban population (thousands)	658,552	678,781	699,892
MDGs			
Access to improved water sources 2010 (%)[a]	91	91	91
Access to improved sanitation 2010 (%)[a]	64	64	64
IBNET sourced data			
Number of utilities reporting in IBNET sample	40	3	3
Population served (water), (thousands)	37,576	5,300	5,590
Size of the sample: Total population living in service area (water supply), (thousands)	35,697	5,035	5,366
Services coverage			
1.1 Water coverage (%)	95	95	96
2.1 Sewerage coverage (%)	38	41	56
Operational efficiency			
13.2 Electrical energy costs vs. operating costs (%) (share of energy cost as % of operational expenses)	13	13	13
6.1 Nonrevenue water (%)	20.00	22.00	21.00
6.2 Nonrevenue water (m^3/km/day)	41	41	37
12.3 Staff W/1,000 W population served (W/1,000 W population served)	1.00	1.00	1.00
15.1 Continuity of service (hrs/day) (duration of water supply, hours)	24.00	24.00	24.00
Financial efficiency			
8.1 Water sold that is metered (%)	100	100	100
23.1 Collection period (days)	87	89	96
23.2 Collection ratio (%)	74	73	77
18.1 Average revenue W & WW (US$/m^3 water sold)	0.29	0.32	0.32
11.1 Operational cost W & WW (US$/m^3 water sold)	0.37	0.40	0.43
24.1 Operating cost coverage (ratio)	0.78	0.80	0.76
Production and consumption			
3.1 Water production (l/person/day)	197.00	200.00	200.00
4.1 Total water consumption (l/person/day)	157.00	156.00	159.00
4.7 Residential consumption (l/person/day)	69	71	73
Poverty and affordability			
19.1 Total revenues/service population/GNI (% GNI per capita) (average revenues)	0.39	0.37	0.32
19.2 Annual bill for households consuming 6 m^3 of water/month (US$/yr)	19.86	20.95	21.50
21.1 Ratio of industrial to residential tariff (level of cross-subsidy)	1.29	1.37	1.28

a. UNICEF and WHO 2012.

IBNET Indicator/Country: Colombia

Latest year available	2008	2009	2010
Surface area (km^2)	1,141,748	1,141,748	1,141,748
GNI per capita, Atlas method (current US$)	4,640	5,030	5,460
Total population (thousands)	45,006	45,654	46,295
Urban population (%)	74	75	75
Total urban population (thousands)	33,504	34,118	34,730
MDGs			
Access to improved water sources 2010 (%)[a]	92	92	92
Access to improved sanitation 2010 (%)[a]	77	77	77
IBNET sourced data			
Number of utilities reporting in IBNET sample	38	38	38
Population served (water), (thousands)	24,090	24,098	25,805
Size of the sample: Total population living in service area (water supply), (thousands)	24,085	25,449	27,479
Services coverage			
1.1 Water coverage (%)	100	95	94
2.1 Sewerage coverage (%)	95	92	90
Operational efficiency			
13.2 Electrical energy costs vs. operating costs (%) (share of energy cost as % of operational expenses)	—	—	—
6.1 Nonrevenue water (%)	—	46.00	79.00
6.2 Nonrevenue water (m^3/km/day)	—	63	75
12.3 Staff W/1,000 W population served (W/1,000 W population served)	—	0.40	0.30
15.1 Continuity of service (hrs/day) (duration of water supply, hours)	24.00	23.70	23.80
Financial efficiency			
8.1 Water sold that is metered (%)	115	94	100
23.1 Collection period (days)	51	36	106
23.2 Collection ratio (%)	81	94	—
18.1 Average revenue W & WW (US$/m^3 water sold)	2.11	1.59	1.78
11.1 Operational cost W & WW (US$/m^3 water sold)	1.38	0.60	1.62
24.1 Operating cost coverage (ratio)	1.53	2.64	1.11
Production and consumption			
3.1 Water production (l/person/day)	—	220.00	216.00
4.1 Total water consumption (l/person/day)	96.00	120.00	109.00
4.7 Residential consumption (l/person/day)	90	96	79
Poverty and affordability			
19.1 Total revenues/service population/GNI (% GNI per capita) (average revenues)	1.59	1.38	1.30
19.2 Annual bill for households consuming 6 m^3 of water/month (US$/yr)	—	—	—
21.1 Ratio of industrial to residential tariff (level of cross-subsidy)	5.82	1.51	1.43

a. UNICEF and WHO 2012.

IBNET Indicator/Country: Democratic Republic of Congo

Latest year available	2003	2004	2005
Surface area (km^2)	2,344,858	2,344,858	2,344,858
GNI per capita, Atlas method (current US$)	100	110	120
Total population (thousands)	54,098	55,755	57,421
Urban population (%)	30	31	31
Total urban population (thousands)	16,455	17,166	17,892
MDGs			
Access to improved water sources 2010 (%)[a]	45	45	45
Access to improved sanitation 2010 (%)[a]	24	24	24
IBNET sourced data			
Number of utilities reporting in IBNET sample	1	1	1
Population served (water), (thousands)	5,166	5,325	5,490
Size of the sample: Total population living in service area (water supply), (thousands)	8,468	8,730	9,000
Services coverage			
1.1 Water coverage (%)	61	61	61
2.1 Sewerage coverage (%)	24	24	24
Operational efficiency			
13.2 Electrical energy costs vs. operating costs (%) (share of energy cost as % of operational expenses)	—	—	—
6.1 Nonrevenue water (%)	44.00	38.00	35.00
6.2 Nonrevenue water (m^3/km/day)	20	17	16
12.3 Staff W/1,000 W population served (W/1,000 W population served)	—	—	—
15.1 Continuity of service (hrs/day) (duration of water supply, hours)	11.00	11.00	11.00
Financial efficiency			
8.1 Water sold that is metered (%)	—	—	—
23.1 Collection period (days)	1,327	2,134	1,834
23.2 Collection ratio (%)	—	—	—
18.1 Average revenue W & WW (US$/m^3 water sold)	0.28	0.34	0.49
11.1 Operational cost W & WW (US$/m^3 water sold)	0.77	1.04	0.76
24.1 Operating cost coverage (ratio)	0.36	0.33	0.64
Production and consumption			
3.1 Water production (l/person/day)	112.50	111.29	104.62
4.1 Total water consumption (l/person/day)	63.00	69.00	68.00
4.7 Residential consumption (l/person/day)	—	—	—
Poverty and affordability			
19.1 Total revenues/service population/GNI (% GNI per capita) (average revenues)	6.44	7.78	10.13
19.2 Annual bill for households consuming 6 m^3 of water/month (US$/yr)	—	—	—
21.1 Ratio of industrial to residential tariff (level of cross-subsidy)	—	—	—

a. UNICEF and WHO 2012.

IBNET Indicator/Country: Costa Rica

Latest year available	2008	2009	2010
Surface area (km²)	51,100	51,100	51,100
GNI per capita, Atlas method (current US$)	6,070	6,140	6,910
Total population (thousands)	4,522	4,591	4,659
Urban population (%)	63	64	64
Total urban population (thousands)	2,858	2,924	2,990
MDGs			
Access to improved water sources 2010 (%)[a]	97	97	97
Access to improved sanitation 2010 (%)[a]	95	95	95
IBNET sourced data			
Number of utilities reporting in IBNET sample	2	2	2
Population served (water), (thousands)	2,094	2,147	2,112
Size of the sample: Total population living in service area (water supply), (thousands)	2,117	2,162	2,112
Services coverage			
1.1 Water coverage (%)	99	99	100
2.1 Sewerage coverage (%)	34	33	33
Operational efficiency			
13.2 Electrical energy costs vs. operating costs (%) (share of energy cost as % of operational expenses)	—	—	—
6.1 Nonrevenue water (%)	50.00	48.00	48.00
6.2 Nonrevenue water (m³/km/day)	53	48	47
12.3 Staff W/1,000 W population served (W/1,000 W population served)	—	—	—
15.1 Continuity of service (hrs/day) (duration of water supply, hours)	24.00	24.00	23.50
Financial efficiency			
8.1 Water sold that is metered (%)	97	98	97
23.1 Collection period (days)	12	24	25
23.2 Collection ratio (%)	108	99	50
18.1 Average revenue W & WW (US$/m³ water sold)	0.71	0.77	1.05
11.1 Operational cost W & WW (US$/m³ water sold)	0.50	0.62	0.77
24.1 Operating cost coverage (ratio)	1.43	1.24	1.36
Production and consumption			
3.1 Water production (l/person/day)	450.00	424.00	429.00
4.1 Total water consumption (l/person/day)	226.00	221.00	225.00
4.7 Residential consumption (l/person/day)	179	173	175
Poverty and affordability			
19.1 Total revenues/service population/GNI (% GNI per capita) (average revenues)	0.96	1.01	1.25
19.2 Annual bill for households consuming 6 m³ of water/month (US$/yr)	—	—	—
21.1 Ratio of industrial to residential tariff (level of cross-subsidy)	3.42	3.01	2.99

a. UNICEF and WHO 2012.

IBNET Indicator/Country: Côte d'Ivoire

Latest year available	2002	2003	2004
Surface area (km²)	322,463	322,463	322,463
GNI per capita, Atlas method (current US$)	580	640	770
Total population (thousands)	17,181	17,456	17,732
Urban population (%)	45	46	46
Total urban population (thousands)	7,707	7,945	8,188
MDGs			
Access to improved water sources 2010 (%)[a]	80	80	80
Access to improved sanitation 2010 (%)[a]	24	24	24
IBNET sourced data			
Number of utilities reporting in IBNET sample	1	1	1
Population served (water), (thousands)	6,234	6,383	6,590
Size of the sample: Total population living in service area (water supply), (thousands)	8,180	8,426	8,678
Services coverage			
1.1 Water coverage (%)	76	76	76
2.1 Sewerage coverage (%)	26	29	26
Operational efficiency			
13.2 Electrical energy costs vs. operating costs (%) (share of energy cost as % of operational expenses)	5	4	5
6.1 Nonrevenue water (%)	19.00	20.00	21.00
6.2 Nonrevenue water (m³/km/day)	7	7	8
12.3 Staff W/1,000 W population served (W/1,000 W population served)	0.20	0.20	0.20
15.1 Continuity of service (hrs/day) (duration of water supply, hours)	24.00	24.00	24.00
Financial efficiency			
8.1 Water sold that is metered (%)	100	100	100
23.1 Collection period (days)	6	2	7
23.2 Collection ratio (%)	95	95	94
18.1 Average revenue W & WW (US$/m³ water sold)	0.50	0.51	0.65
11.1 Operational cost W & WW (US$/m³ water sold)	0.51	0.51	0.63
24.1 Operating cost coverage (ratio)	0.99	1.00	1.04
Production and consumption			
3.1 Water production (l/person/day)	65.43	66.25	67.09
4.1 Total water consumption (l/person/day)	53.00	53.00	53.00
4.7 Residential consumption (l/person/day)	40	40	39
Poverty and affordability			
19.1 Total revenues/service population/GNI (% GNI per capita) (average revenues)	1.67	1.54	1.63
19.2 Annual bill for households consuming 6 m³ of water/month (US$/yr)	24.28	29.11	32.03
21.1 Ratio of industrial to residential tariff (level of cross-subsidy)	1.03	1.03	1.04

a. UNICEF and WHO 2012.

IBNET Indicator/Country: Croatia

Latest year available	2002	2003	2004
Surface area (km²)	56,594	56,594	56,594
GNI per capita, Atlas method (current US$)	5,390	6,390	8,150
Total population (thousands)	11,175	11,208	11,235
Urban population (%)	56	56	56
Total urban population (thousands)	6,254	6,292	6,292
MDGs			
Access to improved water sources 2010 (%)[a]	99	99	99
Access to improved sanitation 2010 (%)[a]	99	99	99
IBNET sourced data			
Number of utilities reporting in IBNET sample	21	21	21
Population served (water), (thousands)	1,747	1,758	1,758
Size of the sample: Total population living in service area (water supply), (thousands)	1,894	1,899	1,903
Services coverage			
1.1 Water coverage (%)	92	93	93
2.1 Sewerage coverage (%)	75	76	76
Operational efficiency			
13.2 Electrical energy costs vs. operating costs (%) (share of energy cost as % of operational expenses)	10	10	10
6.1 Nonrevenue water (%)	20.00	19.00	19.00
6.2 Nonrevenue water (m³/km/day)	13	14	14
12.3 Staff W/1,000 W population served (W/1,000 W population served)	—	—	—
15.1 Continuity of service (hrs/day) (duration of water supply, hours)	24.00	24.00	24.00
Financial efficiency			
8.1 Water sold that is metered (%)	82	82	82
23.1 Collection period (days)	114	93	93
23.2 Collection ratio (%)	71	67	67
18.1 Average revenue W & WW (US$/m³ water sold)	0.52	0.68	0.68
11.1 Operational cost W & WW (US$/m³ water sold)	0.41	0.51	0.51
24.1 Operating cost coverage (ratio)	1.27	1.33	1.33
Production and consumption			
3.1 Water production (l/person/day)	435.00	432.10	432.10
4.1 Total water consumption (l/person/day)	348.00	350.00	350.00
4.7 Residential consumption (l/person/day)	263	266	266
Poverty and affordability			
19.1 Total revenues/service population/GNI (% GNI per capita) (average revenues)	1.23	1.36	1.07
19.2 Annual bill for households consuming 6 m³ of water/month (US$/yr)	—	—	—
21.1 Ratio of industrial to residential tariff (level of cross-subsidy)	4.00	9.92	10.92

a. UNICEF and WHO 2012.

IBNET Indicator/Country: Czech Republic

Latest year available	2008	2009	2010
Surface area (km^2)	78,867	78,867	78,867
GNI per capita, Atlas method (current US$)	17,840	17,920	18,370
Total population (thousands)	10,424	10,487	10,520
Urban population (%)	74	74	73
Total urban population (thousands)	7,683	7,719	7,679
MDGs			
Access to improved water sources 2010 (%)[a]	100	100	100
Access to improved sanitation 2010 (%)[a]	98	98	98
IBNET sourced data			
Number of utilities reporting in IBNET sample	23	23	21
Population served (water), (thousands)	6,632	6,676	6,127
Size of the sample: Total population living in service area (water supply), (thousands)	6,932	6,959	6,377
Services coverage			
1.1 Water coverage (%)	96	96	96
2.1 Sewerage coverage (%)	82	82	82
Operational efficiency			
13.2 Electrical energy costs vs. operating costs (%) (share of energy cost as % of operational expenses)	7	7	7
6.1 Nonrevenue water (%)	20.00	20.00	20.00
6.2 Nonrevenue water (m^3/km/day)	7	7	7
12.3 Staff W/1,000 W population served (W/1,000 W population served)	0.60	0.60	0.60
15.1 Continuity of service (hrs/day) (duration of water supply, hours)	24.00	24.00	24.00
Financial efficiency			
8.1 Water sold that is metered (%)	99	99	100
23.1 Collection period (days)	177	183	185
23.2 Collection ratio (%)	98	97	97
18.1 Average revenue W & WW (US$/m^3 water sold)	2.88	2.41	2.31
11.1 Operational cost W & WW (US$/m^3 water sold)	2.44	2.14	1.80
24.1 Operating cost coverage (ratio)	1.17	1.06	1.25
Production and consumption			
3.1 Water production (l/person/day)	220.00	215.00	209.00
4.1 Total water consumption (l/person/day)	177.00	172.00	167.00
4.7 Residential consumption (l/person/day)	96	94	91
Poverty and affordability			
19.1 Total revenues/service population/GNI (% GNI per capita) (average revenues)	1.04	0.84	0.77
19.2 Annual bill for households consuming 6 m^3 of water/month (US$/yr)	134.81	118.09	106.61
21.1 Ratio of industrial to residential tariff (level of cross-subsidy)	1.02	1.02	1.02

a. UNICEF and WHO 2012.

IBNET Indicator/Country: Ecuador

Latest year available	2008	2009	2010
Surface area (km^2)	256,369	256,369	256,369
GNI per capita, Atlas method (current US$)	3,540	4,070	4,330
Total population (thousands)	14,057	14,262	14,465
Urban population (%)	66	66	67
Total urban population (thousands)	9,216	9,443	9,672
MDGs			
Access to improved water sources 2010 (%)[a]	94	94	94
Access to improved sanitation 2010 (%)[a]	92	92	92
IBNET sourced data			
Number of utilities reporting in IBNET sample	1	2	1
Population served (water), (thousands)	2,061	4,214	2,144
Size of the sample: Total population living in service area (water supply), (thousands)	2,319	4,501	2,243
Services coverage			
1.1 Water coverage (%)	89	94	96
2.1 Sewerage coverage (%)	47	70	90
Operational efficiency			
13.2 Electrical energy costs vs. operating costs (%) (share of energy cost as % of operational expenses)	—	—	3
6.1 Nonrevenue water (%)	65.00	51.00	31.00
6.2 Nonrevenue water (m^3/km/day)	141	78	37
12.3 Staff W/1,000 W population served (W/1,000 W population served)	—	—	0.80
15.1 Continuity of service (hrs/day) (duration of water supply, hours)	—	24.00	24.00
Financial efficiency			
8.1 Water sold that is metered (%)	93	94	93
23.1 Collection period (days)	286	169	59
23.2 Collection ratio (%)	101	110	116
18.1 Average revenue W & WW (US$/m^3 water sold)	0.70	0.63	0.59
11.1 Operational cost W & WW (US$/m^3 water sold)	0.43	0.50	0.37
24.1 Operating cost coverage (ratio)	1.64	1.26	1.61
Production and consumption			
3.1 Water production (l/person/day)	—	374.00	291.00
4.1 Total water consumption (l/person/day)	163.00	184.00	201.00
4.7 Residential consumption (l/person/day)	100	132	162
Poverty and affordability			
19.1 Total revenues/service population/GNI (% GNI per capita) (average revenues)	1.18	1.04	1.00
19.2 Annual bill for households consuming 6 m^3 of water/month (US$/yr)	—	22.32	—
21.1 Ratio of industrial to residential tariff (level of cross-subsidy)	—	1.14	2.22

a. UNICEF and WHO 2012.

IBNET Indicator/Country: El Salvador

Latest year available	2006
Surface area (km²)	21,041
GNI per capita, Atlas method (current US$)	2,990
Total population (thousands)	6,074
Urban population (%)	62
Total urban population (thousands)	3,766
MDGs	
Access to improved water sources 2010 (%)[a]	88
Access to improved sanitation 2010 (%)[a]	87
IBNET sourced data	
Number of utilities reporting in IBNET sample	1
Population served (water), (thousands)	3,951
Size of the sample: Total population living in service area (water supply), (thousands)	5,382
Services coverage	
1.1 Water coverage (%)	73
2.1 Sewerage coverage (%)	39
Operational efficiency	
13.2 Electrical energy costs vs. operating costs (%) (share of energy cost as % of operational expenses)	—
6.1 Nonrevenue water (%)	34.00
6.2 Nonrevenue water (m³/km/day)	75
12.3 Staff W/1,000 W population served (W/1,000 W population served)	0.50
15.1 Continuity of service (hrs/day) (duration of water supply, hours)	24.00
Financial efficiency	
8.1 Water sold that is metered (%)	65
23.1 Collection period (days)	93
23.2 Collection ratio (%)	104
18.1 Average revenue W & WW (US$/m³ water sold)	0.04
11.1 Operational cost W & WW (US$/m³ water sold)	0.03
24.1 Operating cost coverage (ratio)	1.17
Production and consumption	
3.1 Water production (l/person/day)	239.39
4.1 Total water consumption (l/person/day)	158.00
4.7 Residential consumption (l/person/day)	122
Poverty and affordability	
19.1 Total revenues/service population/GNI (% GNI per capita) (average revenues)	0.08
19.2 Annual bill for households consuming 6 m³ of water/month (US$/yr)	—
21.1 Ratio of industrial to residential tariff (level of cross-subsidy)	—

a. UNICEF and WHO 2012.

IBNET Indicator/Country: Arab Republic of Egypt

Latest year available	2008	2009	2010
Surface area (km²)	1,002,000	1,002,000	1,002,000
GNI per capita, Atlas method (current US$)	1,880	2,270	2,550
Total population (thousands)	78,323	79,716	81,121
Urban population (%)	43	43	43
Total urban population (thousands)	33,864	34,521	35,186
MDGs			
Access to improved water sources 2010 (%)[a]	99	99	99
Access to improved sanitation 2010 (%)[a]	95	95	95
IBNET sourced data			
Number of utilities reporting in IBNET sample	21	21	21
Population served (water), (thousands)	83,134	77,530	79,211
Size of the sample: Total population living in service area (water supply), (thousands)	83,838	78,332	80,014
Services coverage			
1.1 Water coverage (%)	99	99	99
2.1 Sewerage coverage (%)	43	47	50
Operational efficiency			
13.2 Electrical energy costs vs. operating costs (%) (share of energy cost as % of operational expenses)	17	17	18
6.1 Nonrevenue water (%)	31.00	28.00	28.00
6.2 Nonrevenue water (m³/km/day)	53	39	39
12.3 Staff W/1,000 W population served (W/1,000 W population served)	0.60	0.80	0.90
15.1 Continuity of service (hrs/day) (duration of water supply, hours)	24.00	24.00	24.00
Financial efficiency			
8.1 Water sold that is metered (%)	61	66	76
23.1 Collection period (days)	530	414	279
23.2 Collection ratio (%)	92	86	84
18.1 Average revenue W & WW (US$/m³ water sold)	0.14	0.16	0.19
11.1 Operational cost W & WW (US$/m³ water sold)	0.15	0.16	0.18
24.1 Operating cost coverage (ratio)	0.98	1.02	1.13
Production and consumption			
3.1 Water production (l/person/day)	237.00	245.00	253.00
4.1 Total water consumption (l/person/day)	164.00	175.00	183.00
4.7 Residential consumption (l/person/day)	119	122	126
Poverty and affordability			
19.1 Total revenues/service population/GNI (% GNI per capita) (average revenues)	0.45	0.45	0.50
19.2 Annual bill for households consuming 6 m³ of water/month (US$/yr)	5.29	5.25	6.02
21.1 Ratio of industrial to residential tariff (level of cross-subsidy)	7.72	4.77	4.47

a. UNICEF and WHO 2012.

IBNET Indicator/Country: Ethiopia

Latest year available	2007	2008	2009
Surface area (km^2)	1,104,300	1,104,300	1,104,300
GNI per capita, Atlas method (current US$)	230	280	320
Total population (thousands)	77,718	79,446	81,188
Urban population (%)	16	16	17
Total urban population (thousands)	12,530	12,977	13,433
MDGs			
Access to improved water sources 2010 (%)[a]	44	44	44
Access to improved sanitation 2010 (%)[a]	21	21	21
IBNET sourced data			
Number of utilities reporting in IBNET sample	6	6	6
Population served (water), (thousands)	3,570	3,696	3,796
Size of the sample: Total population living in service area (water supply), (thousands)	3,643	3,742	3,844
Services coverage			
1.1 Water coverage (%)	98	99	99
2.1 Sewerage coverage (%)	—	—	—
Operational efficiency			
13.2 Electrical energy costs vs. operating costs (%) (share of energy cost as % of operational expenses)	39	36	45
6.1 Nonrevenue water (%)	38.00	41.00	39.00
6.2 Nonrevenue water (m^3/km/day)	39	38	36
12.3 Staff W/1,000 W population served (W/1,000 W population served)	0.40	0.60	0.60
15.1 Continuity of service (hrs/day) (duration of water supply, hours)	16.30	17.00	16.30
Financial efficiency			
8.1 Water sold that is metered (%)	100	100	100
23.1 Collection period (days)	39	34	23
23.2 Collection ratio (%)	102	93	104
18.1 Average revenue W & WW (US$/m^3 water sold)	0.39	0.42	0.31
11.1 Operational cost W & WW (US$/m^3 water sold)	0.14	0.18	0.15
24.1 Operating cost coverage (ratio)	2.75	2.28	2.05
Production and consumption			
3.1 Water production (l/person/day)	76.00	78.00	81.00
4.1 Total water consumption (l/person/day)	47.00	46.00	49.00
4.7 Residential consumption (l/person/day)	32	24	28
Poverty and affordability			
19.1 Total revenues/service population/GNI (% GNI per capita) (average revenues)	2.91	2.52	1.73
19.2 Annual bill for households consuming 6 m^3 of water/month (US$/yr)	—	—	—
21.1 Ratio of industrial to residential tariff (level of cross-subsidy)	3.58	0.41	0.35

a. UNICEF and WHO 2012.

IBNET Indicator/Country: Fiji

Latest year available	2012	2013
Surface area (km²)	18,270	18,270
GNI per capita, Atlas method (current US$)	4,110	4,200
Total population (thousands)	877	885
Urban population (%)	53	53
Total urban population (thousands)	461	469
MDGs		
Access to improved water sources 2010 (%)[a]	98	98
Access to improved sanitation 2010 (%)[a]	83	83
IBNET sourced data		
Number of utilities reporting in IBNET sample	1	1
Population served (water), (thousands)	600	660
Size of the sample: Total population living in service area (water supply), (thousands)	520	600
Services coverage		
1.1 Water coverage (%)	87	98
2.1 Sewerage coverage (%)	37	21
Operational efficiency		
13.2 Electrical energy costs vs. operating costs (%) (share of energy cost as % of operational expenses)	23	38
6.1 Nonrevenue water (%)	50.00	51.00
6.2 Nonrevenue water (m³/km/day)	47	43
12.3 Staff W/1,000 W population served (W/1,000 W population served)	—	—
15.1 Continuity of service (hrs/day) (duration of water supply, hours)	24.00	19.00
Financial efficiency		
8.1 Water sold that is metered (%)	100	100
23.1 Collection period (days)	478	525
23.2 Collection ratio (%)	94	94
18.1 Average revenue W & WW (US$/m³ water sold)	0.26	0.27
11.1 Operational cost W & WW (US$/m³ water sold)	0.63	0.60
24.1 Operating cost coverage (ratio)	0.42	0.45
Production and consumption		
3.1 Water production (l/person/day)	589.00	506.00
4.1 Total water consumption (l/person/day)	294.00	247.00
4.7 Residential consumption (l/person/day)	—	—
Poverty and affordability		
19.1 Total revenues/service population/GNI (% GNI per capita) (average revenues)	0.68	0.58
19.2 Annual bill for households consuming 6 m³ of water/month (US$/yr)	—	210.64
21.1 Ratio of industrial to residential tariff (level of cross-subsidy)	1.00	1.00

a. UNICEF and WHO 2012.

IBNET Indicator/Country: Gabon

Latest year available	2007	2008	2009
Surface area (km^2)	267,668	267,668	267,668
GNI per capita, Atlas method (current US$)	6,470	7,500	7,860
Total population (thousands)	1,424	1,450	1,478
Urban population (%)	84	85	85
Total urban population (thousands)	1,202	1,231	1,261
MDGs			
Access to improved water sources 2010 (%)[a]	87	87	87
Access to improved sanitation 2010 (%)[a]	33	33	33
IBNET sourced data			
Number of utilities reporting in IBNET sample	1	1	1
Population served (water), (thousands)	932	932	933
Size of the sample: Total population living in service area (water supply), (thousands)	1,218	1,218	1,222
Services coverage			
1.1 Water coverage (%)	77	77	76
2.1 Sewerage coverage (%)	—	—	—
Operational efficiency			
13.2 Electrical energy costs vs. operating costs (%) (share of energy cost as % of operational expenses)	36	33	23
6.1 Nonrevenue water (%)	18.00	21.00	23.00
6.2 Nonrevenue water (m^3/km/day)	20	24	27
12.3 Staff W/1,000 W population served (W/1,000 W population served)	1.60	1.60	1.60
15.1 Continuity of service (hrs/day) (duration of water supply, hours)	24.00	24.00	24.00
Financial efficiency			
8.1 Water sold that is metered (%)	100	100	100
23.1 Collection period (days)	85	87	89
23.2 Collection ratio (%)	98	98	98
18.1 Average revenue W & WW (US$/m^3 water sold)	0.66	0.73	0.73
11.1 Operational cost W & WW (US$/m^3 water sold)	0.21	0.25	0.27
24.1 Operating cost coverage (ratio)	3.05	2.92	2.68
Production and consumption			
3.1 Water production (l/person/day)	218.00	228.00	238.00
4.1 Total water consumption (l/person/day)	178.00	181.00	184.00
4.7 Residential consumption (l/person/day)	—	—	—
Poverty and affordability			
19.1 Total revenues/service population/GNI (% GNI per capita) (average revenues)	0.66	0.64	0.62
19.2 Annual bill for households consuming 6 m^3 of water/month (US$/yr)	—	—	—
21.1 Ratio of industrial to residential tariff (level of cross-subsidy)	—	—	—

a. UNICEF and WHO 2012.

IBNET Indicator/Country: The Gambia

Latest year available	2005
Surface area (km^2)	11,295
GNI per capita, Atlas method (current US$)	390
Total population (thousands)	1,504
Urban population (%)	53
Total urban population (thousands)	799
MDGs	
Access to improved water sources 2010 (%)[a]	89
Access to improved sanitation 2010 (%)[a]	68
IBNET sourced data	
Number of utilities reporting in IBNET sample	1
Population served (water), (thousands)	626
Size of the sample: Total population living in service area (water supply), (thousands)	821
Services coverage	
1.1 Water coverage (%)	76
2.1 Sewerage coverage (%)	5
Operational efficiency	
13.2 Electrical energy costs vs. operating costs (%) (share of energy cost as % of operational expenses)	—
6.1 Nonrevenue water (%)	17.00
6.2 Nonrevenue water (m^3/km/day)	25
12.3 Staff W/1,000 W population served (W/1,000 W population served)	—
15.1 Continuity of service (hrs/day) (duration of water supply, hours)	—
Financial efficiency	
8.1 Water sold that is metered (%)	—
23.1 Collection period (days)	—
23.2 Collection ratio (%)	94
18.1 Average revenue W & WW (US$/m^3 water sold)	0.26
11.1 Operational cost W & WW (US$/m^3 water sold)	0.33
24.1 Operating cost coverage (ratio)	0.79
Production and consumption	
3.1 Water production (l/person/day)	68.67
4.1 Total water consumption (l/person/day)	57.00
4.7 Residential consumption (l/person/day)	—
Poverty and affordability	
19.1 Total revenues/service population/GNI (% GNI per capita) (average revenues)	1.39
19.2 Annual bill for households consuming 6 m^3 of water/month (US$/yr)	—
21.1 Ratio of industrial to residential tariff (level of cross-subsidy)	—

a. UNICEF and WHO 2012.

IBNET Indicator/Country: Georgia

Latest year available	2006	2007	2008
Surface area (km^2)	69,700	69,700	69,700
GNI per capita, Atlas method (current US$)	1,680	2,090	2,460
Total population (thousands)	4,398	4,388	4,384
Urban population (%)	53	53	53
Total urban population (thousands)	2,312	2,311	2,312
MDGs			
Access to improved water sources 2010 (%)[a]	98	98	98
Access to improved sanitation 2010 (%)[a]	95	95	95
IBNET sourced data			
Number of utilities reporting in IBNET sample	14	14	14
Population served (water), (thousands)	1,230	1,242	1,260
Size of the sample: Total population living in service area (water supply), (thousands)	1,301	1,303	1,318
Services coverage			
1.1 Water coverage (%)	95	95	96
2.1 Sewerage coverage (%)	84	83	82
Operational efficiency			
13.2 Electrical energy costs vs. operating costs (%) (share of energy cost as % of operational expenses)	14	15	18
6.1 Nonrevenue water (%)	43.00	43.00	46.00
6.2 Nonrevenue water (m^3/km/day)	126	131	129
12.3 Staff W/1,000 W population served (W/1,000 W population served)	2.60	2.50	2.60
15.1 Continuity of service (hrs/day) (duration of water supply, hours)	14.07	14.71	14.71
Financial efficiency			
8.1 Water sold that is metered (%)	8	8	8
23.1 Collection period (days)	207	152	92
23.2 Collection ratio (%)	105	98	114
18.1 Average revenue W & WW (US$/m^3 water sold)	0.09	0.13	0.14
11.1 Operational cost W & WW (US$/m^3 water sold)	0.18	0.14	0.14
24.1 Operating cost coverage (ratio)	0.50	0.95	0.98
Production and consumption			
3.1 Water production (l/person/day)	1,201.75	1,231.58	1,025.93
4.1 Total water consumption (l/person/day)	685.00	702.00	554.00
4.7 Residential consumption (l/person/day)	603	619	616
Poverty and affordability			
19.1 Total revenues/service population/GNI (% GNI per capita) (average revenues)	1.34	1.59	1.15
19.2 Annual bill for households consuming 6 m^3 of water/month (US$/yr)	4.94	7.49	7.77
21.1 Ratio of industrial to residential tariff (level of cross-subsidy)	36.38	47.78	45.44

a. UNICEF and WHO 2012.

IBNET Indicator/Country: Ghana

Latest year available	2007	2008	2009
Surface area (km²)	238,539	238,539	238,539
GNI per capita, Atlas method (current US$)	810	1,160	1,200
Total population (thousands)	22,712	23,264	23,824
Urban population (%)	49	50	51
Total urban population (thousands)	11,152	11,587	12,034
MDGs			
Access to improved water sources 2010 (%)[a]	86	86	86
Access to improved sanitation 2010 (%)[a]	14	14	14
IBNET sourced data			
Number of utilities reporting in IBNET sample	1	1	1
Population served (water), (thousands)	6,249	6,569	6,655
Size of the sample: Total population living in service area (water supply), (thousands)	11,300	11,900	12,100
Services coverage			
1.1 Water coverage (%)	55	55	55
2.1 Sewerage coverage (%)	—	—	—
Operational efficiency			
13.2 Electrical energy costs vs. operating costs (%) (share of energy cost as % of operational expenses)	—	—	—
6.1 Nonrevenue water (%)	52.00	52.00	52.00
6.2 Nonrevenue water (m³/km/day)	40	41	42
12.3 Staff W/1,000 W population served (W/1,000 W population served)	—	—	—
15.1 Continuity of service (hrs/day) (duration of water supply, hours)	24.00	24.00	24.00
Financial efficiency			
8.1 Water sold that is metered (%)	—	—	—
23.1 Collection period (days)	363	369	372
23.2 Collection ratio (%)	90	91	79
18.1 Average revenue W & WW (US$/m³ water sold)	0.74	0.87	0.63
11.1 Operational cost W & WW (US$/m³ water sold)	0.61	0.76	0.54
24.1 Operating cost coverage (ratio)	1.21	1.15	1.16
Production and consumption			
3.1 Water production (l/person/day)	96.00	93.00	95.00
4.1 Total water consumption (l/person/day)	46.00	45.00	46.00
4.7 Residential consumption (l/person/day)	—	—	—
Poverty and affordability			
19.1 Total revenues/service population/GNI (% GNI per capita) (average revenues)	1.53	1.23	0.88
19.2 Annual bill for households consuming 6 m³ of water/month (US$/yr)	—	—	—
21.1 Ratio of industrial to residential tariff (level of cross-subsidy)	1.00	1.00	1.00

a. UNICEF and WHO 2012.

IBNET Indicator/Country: Guinea

Latest year available	2007	2008	2009
Surface area (km^2)	245,857	245,857	245,857
GNI per capita, Atlas method (current US$)	330	340	390
Total population (thousands)	9,374	9,559	9,761
Urban population (%)	34	34	35
Total urban population (thousands)	3,158	3,261	3,372
MDGs			
Access to improved water sources 2010 (%)[a]	74	74	74
Access to improved sanitation 2010 (%)[a]	18	18	18
IBNET sourced data			
Number of utilities reporting in IBNET sample	1	1	1
Population served (water), (thousands)	3,243	3,401	3,530
Size of the sample: Total population living in service area (water supply), (thousands)	4,580	4,731	4,920
Services coverage			
1.1 Water coverage (%)	71	72	72
2.1 Sewerage coverage (%)	—	—	—
Operational efficiency			
13.2 Electrical energy costs vs. operating costs (%) (share of energy cost as % of operational expenses)	6	10	9
6.1 Nonrevenue water (%)	50.00	46.00	43.00
6.2 Nonrevenue water (m^3/km/day)	11	13	13
12.3 Staff W/1,000 W population served (W/1,000 W population served)	—	—	—
15.1 Continuity of service (hrs/day) (duration of water supply, hours)	8.00	8.00	10.00
Financial efficiency			
8.1 Water sold that is metered (%)	89	92	93
23.1 Collection period (days)	—	—	—
23.2 Collection ratio (%)	77	64	72
18.1 Average revenue W & WW (US$/m^3 water sold)	0.56	0.59	0.65
11.1 Operational cost W & WW (US$/m^3 water sold)	1.07	0.75	0.68
24.1 Operating cost coverage (ratio)	0.52	0.79	0.96
Production and consumption			
3.1 Water production (l/person/day)	30.00	37.04	40.35
4.1 Total water consumption (l/person/day)	15.00	20.00	23.00
4.7 Residential consumption (l/person/day)	8	9	9
Poverty and affordability			
19.1 Total revenues/service population/GNI (% GNI per capita) (average revenues)	0.93	1.27	1.40
19.2 Annual bill for households consuming 6 m^3 of water/month (US$/yr)	—	—	—
21.1 Ratio of industrial to residential tariff (level of cross-subsidy)	21.23	32.08	39.28

a. UNICEF and WHO 2012.

IBNET Indicator/Country: Hungary

Latest year available	2006	2007
Surface area (km²)	93,028	93,028
GNI per capita, Atlas method (current US$)	11,040	11,510
Total population (thousands)	10,071	10,056
Urban population (%)	67	67
Total urban population (thousands)	6,718	6,747
MDGs		
Access to improved water sources 2010 (%)[a]	100	100
Access to improved sanitation 2010 (%)[a]	100	100
IBNET sourced data		
Number of utilities reporting in IBNET sample	20	20
Population served (water), (thousands)	4,853	4,853
Size of the sample: Total population living in service area (water supply), (thousands)	4,902	4,902
Services coverage		
1.1 Water coverage (%)	99	99
2.1 Sewerage coverage (%)	70	70
Operational efficiency		
13.2 Electrical energy costs vs. operating costs (%) (share of energy cost as % of operational expenses)	10	11
6.1 Nonrevenue water (%)	41.00	51.00
6.2 Nonrevenue water (m³/km/day)	15	14
12.3 Staff W/1,000 W population served (W/1,000 W population served)	0.90	0.90
15.1 Continuity of service (hrs/day) (duration of water supply, hours)	24.00	24.00
Financial efficiency		
8.1 Water sold that is metered (%)	100	100
23.1 Collection period (days)	45	49
23.2 Collection ratio (%)	94	101
18.1 Average revenue W & WW (US$/m³ water sold)	1.37	1.64
11.1 Operational cost W & WW (US$/m³ water sold)	1.33	1.51
24.1 Operating cost coverage (ratio)	1.03	1.09
Production and consumption		
3.1 Water production (l/person/day)	233.90	238.78
4.1 Total water consumption (l/person/day)	138.00	117.00
4.7 Residential consumption (l/person/day)	109	110
Poverty and affordability		
19.1 Total revenues/service population/GNI (% GNI per capita) (average revenues)	0.63	0.61
19.2 Annual bill for households consuming 6 m³ of water/month (US$/yr)	77.95	99.80
21.1 Ratio of industrial to residential tariff (level of cross-subsidy)	1.32	1.22

a. UNICEF and WHO 2012.

IBNET Indicator/Country: India

Latest year available	2009
Surface area (km²)	3,287,240
GNI per capita, Atlas method (current US$)	1,170
Total population (thousands)	1,207,740
Urban population (%)	31
Total urban population (thousands)	374,400
MDGs	
Access to improved water sources 2010 (%)[a]	92
Access to improved sanitation 2010 (%)[a]	34
IBNET sourced data	
Number of utilities reporting in IBNET sample	28
Population served (water), (thousands)	57,401
Size of the sample: Total population living in service area (water supply), (thousands)	56,838
Services coverage	
1.1 Water coverage (%)	100
2.1 Sewerage coverage (%)	23
Operational efficiency	
13.2 Electrical energy costs vs. operating costs (%) (share of energy cost as % of operational expenses)	43
6.1 Nonrevenue water (%)	41.00
6.2 Nonrevenue water (m³/km/day)	119
12.3 Staff W/1,000 W population served (W/1,000 W population served)	0.90
15.1 Continuity of service (hrs/day) (duration of water supply, hours)	5.20
Financial efficiency	
8.1 Water sold that is metered (%)	39
23.1 Collection period (days)	—
23.2 Collection ratio (%)	82
18.1 Average revenue W & WW (US$/m³ water sold)	0.15
11.1 Operational cost W & WW (US$/m³ water sold)	0.28
24.1 Operating cost coverage (ratio)	0.55
Production and consumption	
3.1 Water production (l/person/day)	194.00
4.1 Total water consumption (l/person/day)	114.00
4.7 Residential consumption (l/person/day)	83
Poverty and affordability	
19.1 Total revenues/service population/GNI (% GNI per capita) (average revenues)	0.53
19.2 Annual bill for households consuming 6 m³ of water/month (US$/yr)	—
21.1 Ratio of industrial to residential tariff (level of cross-subsidy)	9.32

a. UNICEF and WHO 2012.

IBNET Indicator/Country: Indonesia

Latest year available	2002	2003	2004
Surface area (km²)	1,860,360	1,860,360	1,860,360
GNI per capita, Atlas method (current US$)	720	890	1,070
Total population (thousands)	219,026	221,839	224,607
Urban population (%)	44	44	45
Total urban population (thousands)	95,443	98,415	101,410
MDGs			
Access to improved water sources 2010 (%)[a]	82	82	82
Access to improved sanitation 2010 (%)[a]	54	54	54
IBNET sourced data			
Number of utilities reporting in IBNET sample	14	14	7
Population served (water), (thousands)	4,729	5,308	1,952
Size of the sample: Total population living in service area (water supply), (thousands)	10,530	10,874	2,571
Services coverage			
1.1 Water coverage (%)	45	49	76
2.1 Sewerage coverage (%)	11	12	15
Operational efficiency			
13.2 Electrical energy costs vs. operating costs (%) (share of energy cost as % of operational expenses)	17	17	17
6.1 Nonrevenue water (%)	30.00	30.00	30.00
6.2 Nonrevenue water (m³/km/day)	36	37	28
12.3 Staff W/1,000 W population served (W/1,000 W population served)	1.10	1.00	1.00
15.1 Continuity of service (hrs/day) (duration of water supply, hours)	19.14	19.79	19.86
Financial efficiency			
8.1 Water sold that is metered (%)	100	100	100
23.1 Collection period (days)	46	55	56
23.2 Collection ratio (%)	111	110	110
18.1 Average revenue W & WW (US$/m³ water sold)	0.14	0.18	0.20
11.1 Operational cost W & WW (US$/m³ water sold)	0.12	0.15	0.15
24.1 Operating cost coverage (ratio)	1.22	1.21	1.39
Production and consumption			
3.1 Water production (l/person/day)	215.71	202.86	185.71
4.1 Total water consumption (l/person/day)	151.00	142.00	130.00
4.7 Residential consumption (l/person/day)	133	123	117
Poverty and affordability			
19.1 Total revenues/service population/GNI (% GNI per capita) (average revenues)	1.07	1.05	0.89
19.2 Annual bill for households consuming 6 m³ of water/month (US$/yr)	—	—	—
21.1 Ratio of industrial to residential tariff (level of cross-subsidy)	—	—	—

a. UNICEF and WHO 2012.

IBNET Indicator/Country: Jordan

Latest year available	2008	2009	2010
Surface area (km²)	89,342	89,342	89,342
GNI per capita, Atlas method (current US$)	3,530	3,900	4,140
Total population (thousands)	5,787	5,915	6,047
Urban population (%)	82	82	82
Total urban population (thousands)	4,743	4,863	4,987
MDGs			
Access to improved water sources 2010 (%)[a]	97	97	97
Access to improved sanitation 2010 (%)[a]	98	98	98
IBNET sourced data			
Number of utilities reporting in IBNET sample	3	4	4
Population served (water), (thousands)	4,335	6,087	6,268
Size of the sample: Total population living in service area (water supply), (thousands)	4,346	6,109	6,264
Services coverage			
1.1 Water coverage (%)	100	100	100
2.1 Sewerage coverage (%)	73	71	73
Operational efficiency			
13.2 Electrical energy costs vs. operating costs (%) (share of energy cost as % of operational expenses)	33	33	34
6.1 Nonrevenue water (%)	43.00	42.00	36.00
6.2 Nonrevenue water (m³/km/day)	16	15	15
12.3 Staff W/1,000 W population served (W/1,000 W population served)	0.80	0.80	0.80
15.1 Continuity of service (hrs/day) (duration of water supply, hours)	9.00	8.80	8.80
Financial efficiency			
8.1 Water sold that is metered (%)	94	94	94
23.1 Collection period (days)	856	918	683
23.2 Collection ratio (%)	115	121	109
18.1 Average revenue W & WW (US$/m³ water sold)	0.73	0.71	0.69
11.1 Operational cost W & WW (US$/m³ water sold)	0.85	0.80	0.70
24.1 Operating cost coverage (ratio)	0.85	0.89	0.99
Production and consumption			
3.1 Water production (l/person/day)	164.00	150.00	152.00
4.1 Total water consumption (l/person/day)	94.00	86.00	97.00
4.7 Residential consumption (l/person/day)	63	64	73
Poverty and affordability			
19.1 Total revenues/service population/GNI (% GNI per capita) (average revenues)	0.71	0.57	0.59
19.2 Annual bill for households consuming 6 m³ of water/month (US$/yr)	25.97	25.97	25.97
21.1 Ratio of industrial to residential tariff (level of cross-subsidy)	2.46	2.55	2.51

a. UNICEF and WHO 2012.

IBNET Indicator/Country: Kazakhstan

Latest year available	2008	2009	2010
Surface area (km^2)	2,724,900	2,724,900	2,724,900
GNI per capita, Atlas method (current US$)	6,150	6,790	7,440
Total population (thousands)	15,674	16,093	16,323
Urban population (%)	54	54	54
Total urban population (thousands)	8,484	8,679	8,771
MDGs			
Access to improved water sources 2010 (%)[a]	95	95	95
Access to improved sanitation 2010 (%)[a]	97	97	97
IBNET sourced data			
Number of utilities reporting in IBNET sample	24	24	25
Population served (water), (thousands)	5,072	5,125	5,306
Size of the sample: Total population living in service area (water supply), (thousands)	7,592	7,736	7,838
Services coverage			
1.1 Water coverage (%)	81	79	81
2.1 Sewerage coverage (%)	61	62	63
Operational efficiency			
13.2 Electrical energy costs vs. operating costs (%) (share of energy cost as % of operational expenses)	33	29	29
6.1 Nonrevenue water (%)	34.00	35.00	31.00
6.2 Nonrevenue water (m^3/km/day)	82	78	59
12.3 Staff W/1,000 W population served (W/1,000 W population served)	1.40	1.50	1.50
15.1 Continuity of service (hrs/day) (duration of water supply, hours)	24.00	24.00	24.00
Financial efficiency			
8.1 Water sold that is metered (%)	42	46	47
23.1 Collection period (days)	66	130	163
23.2 Collection ratio (%)	101	85	92
18.1 Average revenue W & WW (US$/m^3 water sold)	0.20	0.27	0.34
11.1 Operational cost W & WW (US$/m^3 water sold)	0.25	0.28	0.33
24.1 Operating cost coverage (ratio)	0.80	0.97	1.02
Production and consumption			
3.1 Water production (l/person/day)	485.00	471.00	416.00
4.1 Total water consumption (l/person/day)	318.00	304.00	286.00
4.7 Residential consumption (l/person/day)	130	132	129
Poverty and affordability			
19.1 Total revenues/service population/GNI (% GNI per capita) (average revenues)	0.38	0.44	0.48
19.2 Annual bill for households consuming 6 m^3 of water/month (US$/yr)	30.05	30.99	37.83
21.1 Ratio of industrial to residential tariff (level of cross-subsidy)	0.90	1.54	1.74

a. UNICEF and WHO 2012.

IBNET Indicator/Country: Kenya

Latest year available	2009	2010
Surface area (km^2)	580,367	580,367
GNI per capita, Atlas method (current US$)	780	800
Total population (thousands)	39,462	40,513
Urban population (%)	23	24
Total urban population (thousands)	9,152	9,549
MDGs		
Access to improved water sources 2010 (%)[a]	59	59
Access to improved sanitation 2010 (%)[a]	32	32
IBNET sourced data		
Number of utilities reporting in IBNET sample	48	62
Population served (water), (thousands)	7,231	8,111
Size of the sample: Total population living in service area (water supply), (thousands)	15,025	20,469
Services coverage		
1.1 Water coverage (%)	48	39
2.1 Sewerage coverage (%)	15	15
Operational efficiency		
13.2 Electrical energy costs vs. operating costs (%) (share of energy cost as % of operational expenses)	—	—
6.1 Nonrevenue water (%)	43.00	45.00
6.2 Nonrevenue water (m^3/km/day)	108	103
12.3 Staff W/1,000 W population served (W/1,000 W population served)	0.70	0.70
15.1 Continuity of service (hrs/day) (duration of water supply, hours)	11.00	13.60
Financial efficiency		
8.1 Water sold that is metered (%)	—	—
23.1 Collection period (days)	64	4
23.2 Collection ratio (%)	84	82
18.1 Average revenue W & WW (US$/m^3 water sold)	0.57	0.72
11.1 Operational cost W & WW (US$/m^3 water sold)	0.60	0.75
24.1 Operating cost coverage (ratio)	0.93	0.95
Production and consumption		
3.1 Water production (l/person/day)	110.00	102.00
4.1 Total water consumption (l/person/day)	62.00	56.00
4.7 Residential consumption (l/person/day)	—	36
Poverty and affordability		
19.1 Total revenues/service population/GNI (% GNI per capita) (average revenues)	1.65	1.84
19.2 Annual bill for households consuming 6 m^3 of water/month (US$/yr)	—	—
21.1 Ratio of industrial to residential tariff (level of cross-subsidy)	—	—

a. UNICEF and WHO 2012.

IBNET Indicator/Country: Kiribati

Latest year available	2011	2012	2013
Surface area (km^2)	811	811	811
GNI per capita, Atlas method (current US$)	2,100	2,520	2,600
Total population (thousands)	101	102	104
Urban population (%)	44	44	44
Total urban population (thousands)	44	45	46
MDGs			
Access to improved water sources 2010 (%)[a]	77 (2000)	77 (2000)	77 (2000)
Access to improved sanitation 2010 (%)[a]	44	44	44
IBNET sourced data			
Number of utilities reporting in IBNET sample	1	1	1
Population served (water), (thousands)	50	50	51
Size of the sample: Total population living in service area (water supply), (thousands)	30	31	34
Services coverage			
1.1 Water coverage (%)	60	62	67
2.1 Sewerage coverage (%)	53	26	32
Operational efficiency			
13.2 Electrical energy costs vs. operating costs (%) (share of energy cost as % of operational expenses)	32	—	—
6.1 Nonrevenue water (%)	31.00	75.00	81.00
6.2 Nonrevenue water (m^3/km/day)	4	10	11
12.3 Staff W/1,000 W population served (W/1,000 W population served)	—	—	—
15.1 Continuity of service (hrs/day) (duration of water supply, hours)	4.00	2.00	2.00
Financial efficiency			
8.1 Water sold that is metered (%)	—	22	100
23.1 Collection period (days)	177	1,812	1,896
23.2 Collection ratio (%)	14	95	23
18.1 Average revenue W & WW (US$/m^3 water sold)	2.22	4.32	6.00
11.1 Operational cost W & WW (US$/m^3 water sold)	1.25	6.33	7.07
24.1 Operating cost coverage (ratio)	1.78	0.68	0.85
Production and consumption			
3.1 Water production (l/person/day)	54.00	62.00	58.00
4.1 Total water consumption (l/person/day)	37.00	16.00	11.00
4.7 Residential consumption (l/person/day)	—	—	—
Poverty and affordability			
19.1 Total revenues/service population/GNI (% GNI per capita) (average revenues)	1.43	1.00	0.93
19.2 Annual bill for households consuming 6 m^3 of water/month (US$/yr)	119.92	—	—
21.1 Ratio of industrial to residential tariff (level of cross-subsidy)	—	—	—

a. UNICEF and WHO 2012.

IBNET Indicator/Country: Kuwait

Latest year available	2008	2009	2010
Surface area (km^2)	17,820	17,820	17,820
GNI per capita, Atlas method (current US$)	54,540	45,480	44,100
Total population (thousands)	2,548	2,646	2,737
Urban population (%)	98	98	98
Total urban population (thousands)	2,503	2,599	2,689
MDGs			
Access to improved water sources 2010 (%)[a]	99	99	99
Access to improved sanitation 2010 (%)[a]	100	100	100
IBNET sourced data			
Number of utilities reporting in IBNET sample	1	1	1
Population served (water), (thousands)	3,400	3,442	3,485
Size of the sample: Total population living in service area (water supply), (thousands)	3,400	3,442	3,485
Services coverage			
1.1 Water coverage (%)	100	100	100
2.1 Sewerage coverage (%)	100	100	100
Operational efficiency			
13.2 Electrical energy costs vs. operating costs (%) (share of energy cost as % of operational expenses)	91	91	92
6.1 Nonrevenue water (%)	—	—	—
6.2 Nonrevenue water (m^3/km/day)	—	—	—
12.3 Staff W/1,000 W population served (W/1,000 W population served)	0.70	0.90	1.30
15.1 Continuity of service (hrs/day) (duration of water supply, hours)	24.00	24.00	24.00
Financial efficiency			
8.1 Water sold that is metered (%)	62	30	35
23.1 Collection period (days)	—	—	—
23.2 Collection ratio (%)	100	100	100
18.1 Average revenue W & WW (US$/m^3 water sold)	0.38	0.18	0.30
11.1 Operational cost W & WW (US$/m^3 water sold)	4.20	1.06	2.84
24.1 Operating cost coverage (ratio)	0.09	0.09	0.11
Production and consumption			
3.1 Water production (l/person/day)	1,116.00	2,297.00	1,976.00
4.1 Total water consumption (l/person/day)	1,116.00	2,297.00	1,976.00
4.7 Residential consumption (l/person/day)	538	532	502
Poverty and affordability			
19.1 Total revenues/service population/GNI (% GNI per capita) (average revenues)	0.28	0.33	0.49
19.2 Annual bill for households consuming 6 m^3 of water/month (US$/yr)	48.20	44.32	43.71
21.1 Ratio of industrial to residential tariff (level of cross-subsidy)	54.49	26.67	15.55

a. UNICEF and WHO 2012.

IBNET Indicator/Country: Kyrgyz Republic

Latest year available	2006	2010	2011
Surface area (km^2)	199,951	199,951	199,951
GNI per capita, Atlas method (current US$)	500	840	900
Total population (thousands)	5,218	5,268	5,319
Urban population (%)	35	35	35
Total urban population (thousands)	1,826	1,844	1,862
MDGs			
Access to improved water sources 2010 (%)[a]	90	90	90
Access to improved sanitation 2010 (%)[a]	93	93	93
IBNET sourced data			
Number of utilities reporting in IBNET sample	9	5	5
Population served (water), (thousands)	1,376	1,333	1,387
Size of the sample: Total population living in service area (water supply), (thousands)	669	1,082	1,124
Services coverage			
1.1 Water coverage (%)	56	69	68
2.1 Sewerage coverage (%)	15	23	24
Operational efficiency			
13.2 Electrical energy costs vs. operating costs (%) (share of energy cost as % of operational expenses)	36	26	22
6.1 Nonrevenue water (%)	70.00	76.00	77.00
6.2 Nonrevenue water (m^3/km/day)	83	196	198
12.3 Staff W/1,000 W population served (W/1,000 W population served)	1.80	1.50	1.50
15.1 Continuity of service (hrs/day) (duration of water supply, hours)	23.60	24.00	24.00
Financial efficiency			
8.1 Water sold that is metered (%)	10	31	33
23.1 Collection period (days)	514	167	169
23.2 Collection ratio (%)	58	95	99
18.1 Average revenue W & WW (US$/m^3 water sold)	0.11	0.12	0.15
11.1 Operational cost W & WW (US$/m^3 water sold)	0.11	0.11	0.12
24.1 Operating cost coverage (ratio)	1.03	1.09	1.21
Production and consumption			
3.1 Water production (l/person/day)	453.00	747.00	771.00
4.1 Total water consumption (l/person/day)	137.00	176.00	180.00
4.7 Residential consumption (l/person/day)	64	122	121
Poverty and affordability			
19.1 Total revenues/service population/GNI (% GNI per capita) (average revenues)	1.10	0.92	1.10
19.2 Annual bill for households consuming 6 m^3 of water/month (US$/yr)	6.60	11.96	12.27
21.1 Ratio of industrial to residential tariff (level of cross-subsidy)	1.97	5.77	3.78

a. UNICEF and WHO 2012.

IBNET Indicator/Country: Lao People's Democratic Republic

Latest year available	2006	2007	2008
Surface area (km²)	236,800	236,800	236,800
GNI per capita, Atlas method (current US$)	510	620	760
Total population (thousands)	5,842	5,931	6,022
Urban population (%)	29	30	31
Total urban population (thousands)	1,668	1,763	1,860
MDGs			
Access to improved water sources 2010 (%)[a]	67	67	67
Access to improved sanitation 2010 (%)[a]	63	63	63
IBNET sourced data			
Number of utilities reporting in IBNET sample	2	10	2
Population served (water), (thousands)	321	596	57
Size of the sample: Total population living in service area (water supply), (thousands)	658	1,055	66
Services coverage			
1.1 Water coverage (%)	49	56	87
2.1 Sewerage coverage (%)	—	—	—
Operational efficiency			
13.2 Electrical energy costs vs. operating costs (%) (share of energy cost as % of operational expenses)	15	11	16
6.1 Nonrevenue water (%)	27.00	26.00	21.00
6.2 Nonrevenue water (m³/km/day)	47	26	46
12.3 Staff W/1,000 W population served (W/1,000 W population served)	1.60	1.60	2.20
15.1 Continuity of service (hrs/day) (duration of water supply, hours)	24.00	24.00	24.00
Financial efficiency			
8.1 Water sold that is metered (%)	100	100	100
23.1 Collection period (days)	77	101	49
23.2 Collection ratio (%)	—	—	—
18.1 Average revenue W & WW (US$/m³ water sold)	0.09	0.13	0.15
11.1 Operational cost W & WW (US$/m³ water sold)	0.15	0.24	0.14
24.1 Operating cost coverage (ratio)	0.58	0.54	1.07
Production and consumption			
3.1 Water production (l/person/day)	375.34	336.49	316.46
4.1 Total water consumption (l/person/day)	274.00	249.00	250.00
4.7 Residential consumption (l/person/day)	223	197	154
Poverty and affordability			
19.1 Total revenues/service population/GNI (% GNI per capita) (average revenues)	1.76	1.91	1.80
19.2 Annual bill for households consuming 6 m³ of water/month (US$/yr)	—	—	—
21.1 Ratio of industrial to residential tariff (level of cross-subsidy)	4.02	3.45	2.12

a. UNICEF and WHO 2012.

IBNET Indicator/Country: Lesotho

Latest year available	2006	2007	2008
Surface area (km^2)	30,355	30,355	30,355
GNI per capita, Atlas method (current US$)	910	950	1,050
Total population (thousands)	2,086	2,106	2,127
Urban population (%)	24	25	25
Total urban population (thousands)	501	521	542
MDGs			
Access to improved water sources 2010 (%)[a]	78	78	78
Access to improved sanitation 2010 (%)[a]	26	26	26
IBNET sourced data			
Number of utilities reporting in IBNET sample	1	1	1
Population served (water), (thousands)	259	300	394
Size of the sample: Total population living in service area (water supply), (thousands)	550	560	571
Services coverage			
1.1 Water coverage (%)	47	54	69
2.1 Sewerage coverage (%)	9	13	16
Operational efficiency			
13.2 Electrical energy costs vs. operating costs (%) (share of energy cost as % of operational expenses)	—	—	—
6.1 Nonrevenue water (%)	28.00	30.00	28.00
6.2 Nonrevenue water (m^3/km/day)	39	43	40
12.3 Staff W/1,000 W population served (W/1,000 W population served)	—	—	—
15.1 Continuity of service (hrs/day) (duration of water supply, hours)	24.00	24.00	24.00
Financial efficiency			
8.1 Water sold that is metered (%)	—	—	—
23.1 Collection period (days)	210	226	234
23.2 Collection ratio (%)	—	—	—
18.1 Average revenue W & WW (US$/m^3 water sold)	0.80	0.96	0.88
11.1 Operational cost W & WW (US$/m^3 water sold)	0.71	0.79	0.85
24.1 Operating cost coverage (ratio)	1.13	1.21	1.04
Production and consumption			
3.1 Water production (l/person/day)	161.11	141.43	106.94
4.1 Total water consumption (l/person/day)	116.00	99.00	77.00
4.7 Residential consumption (l/person/day)	—	—	—
Poverty and affordability			
19.1 Total revenues/service population/GNI (% GNI per capita) (average revenues)	3.72	3.65	2.36
19.2 Annual bill for households consuming 6 m^3 of water/month (US$/yr)	—	—	—
21.1 Ratio of industrial to residential tariff (level of cross-subsidy)	—	—	—

a. UNICEF and WHO 2012.

IBNET Indicator/Country: Liberia

Latest year available	2004	2005	2006
Surface area (km^2)	111,369	111,369	111,369
GNI per capita, Atlas method (current US$)	90	120	140
Total population (thousands)	3,093	3,183	3,314
Urban population (%)	46	46	46
Total urban population (thousands)	1,414	1,466	1,538
MDGs			
Access to improved water sources 2010 (%)[a]	73	73	73
Access to improved sanitation 2010 (%)[a]	18	18	18
IBNET sourced data			
Number of utilities reporting in IBNET sample	1	1	1
Population served (water), (thousands)	350	350	350
Size of the sample: Total population living in service area (water supply), (thousands)	1,500	1,500	1,200
Services coverage			
1.1 Water coverage (%)	23	23	29
2.1 Sewerage coverage (%)	10	10	17
Operational efficiency			
13.2 Electrical energy costs vs. operating costs (%) (share of energy cost as % of operational expenses)	—	—	—
6.1 Nonrevenue water (%)	7.00	29.00	49.00
6.2 Nonrevenue water (m^3/km/day)	1	4	10
12.3 Staff W/1,000 W population served (W/1,000 W population served)	0.10	0.20	0.20
15.1 Continuity of service (hrs/day) (duration of water supply, hours)	6.00	6.00	12.00
Financial efficiency			
8.1 Water sold that is metered (%)	—	—	95
23.1 Collection period (days)	80	133	127
23.2 Collection ratio (%)	57	63	75
18.1 Average revenue W & WW (US$/m^3 water sold)	1.15	1.15	1.22
11.1 Operational cost W & WW (US$/m^3 water sold)	0.91	1.17	1.17
24.1 Operating cost coverage (ratio)	1.26	0.98	1.05
Production and consumption			
3.1 Water production (l/person/day)	7.53	8.45	11.76
4.1 Total water consumption (l/person/day)	7.00	6.00	6.00
4.7 Residential consumption (l/person/day)	1	1	1
Poverty and affordability			
19.1 Total revenues/service population/GNI (% GNI per capita) (average revenues)	3.26	2.10	1.91
19.2 Annual bill for households consuming 6 m^3 of water/month (US$/yr)	44.59	45.33	48.00
21.1 Ratio of industrial to residential tariff (level of cross-subsidy)	4.15	3.40	2.54

a. UNICEF and WHO 2012.

IBNET Indicator/Country: Lithuania

Latest year available	2008	2009	2010
Surface area (km^2)	65,200	65,200	65,200
GNI per capita, Atlas method (current US$)	12,000	11,700	11,620
Total population (thousands)	3,358	3,339	3,287
Urban population (%)	67	67	67
Total urban population (thousands)	2,250	2,237	2,202
MDGs			
Access to improved water sources 2010 (%)[a]	98	98	98
Access to improved sanitation 2010 (%)[a]	95	95	95
IBNET sourced data			
Number of utilities reporting in IBNET sample	27	26	26
Population served (water), (thousands)	1,453	1,462	1,449
Size of the sample: Total population living in service area (water supply), (thousands)	1,813	1,821	1,788
Services coverage			
1.1 Water coverage (%)	80	80	82
2.1 Sewerage coverage (%)	73	74	76
Operational efficiency			
13.2 Electrical energy costs vs. operating costs (%) (share of energy cost as % of operational expenses)	15	16	19
6.1 Nonrevenue water (%)	23.00	24.00	24.00
6.2 Nonrevenue water (m^3/km/day)	8	8	7
12.3 Staff W/1,000 W population served (W/1,000 W population served)	0.90	0.80	0.80
15.1 Continuity of service (hrs/day) (duration of water supply, hours)	24.00	24.00	24.00
Financial efficiency			
8.1 Water sold that is metered (%)	99	99	99
23.1 Collection period (days)	77	119	115
23.2 Collection ratio (%)	103	103	105
18.1 Average revenue W & WW (US$/m^3 water sold)	1.91	1.78	1.58
11.1 Operational cost W & WW (US$/m^3 water sold)	1.49	1.42	1.28
24.1 Operating cost coverage (ratio)	1.29	1.22	1.22
Production and consumption			
3.1 Water production (l/person/day)	164.00	152.00	150.00
4.1 Total water consumption (l/person/day)	126.00	115.00	113.00
4.7 Residential consumption (l/person/day)	79	74	72
Poverty and affordability			
19.1 Total revenues/service population/GNI (% GNI per capita) (average revenues)	0.73	0.64	0.56
19.2 Annual bill for households consuming 6 m^3 of water/month (US$/yr)	161.29	151.80	136.72
21.1 Ratio of industrial to residential tariff (level of cross-subsidy)	2.07	2.26	2.23

a. UNICEF and WHO 2012.

IBNET Indicator/Country: Former Yugoslav Republic of Macedonia

Latest year available	2010	2011	2012
Surface area (km^2)	25,713	25,713	25,713
GNI per capita, Atlas method (current US$)	4,580	4,710	4,620
Total population (thousands)	2,061	2,064	2,068
Urban population (%)	59	59	59
Total urban population (thousands)	1,220	1,224	1,228
MDGs			
Access to improved water sources 2010 (%)[a]	100	100	100
Access to improved sanitation 2010 (%)[a]	59	59	59
IBNET sourced data			
Number of utilities reporting in IBNET sample	13	27	24
Population served (water), (thousands)	746	750	
Size of the sample: Total population living in service area (water supply), (thousands)	781	785	
Services coverage			
1.1 Water coverage (%)	96	95	91
2.1 Sewerage coverage (%)	84	84	73
Operational efficiency			
13.2 Electrical energy costs vs. operating costs (%) (share of energy cost as % of operational expenses)	14	13	9
6.1 Nonrevenue water (%)	60.00	60.00	64.00
6.2 Nonrevenue water (m^3/km/day)	114	109	103
12.3 Staff W/1,000 W population served (W/1,000 W population served)	1.10	1.00	1.00
15.1 Continuity of service (hrs/day) (duration of water supply, hours)	24.00	24.00	23.80
Financial efficiency			
8.1 Water sold that is metered (%)	93	94	96
23.1 Collection period (days)	501	487	730
23.2 Collection ratio (%)	96	97	87
18.1 Average revenue W & WW (US$/m^3 water sold)	0.61	0.71	0.82
11.1 Operational cost W & WW (US$/m^3 water sold)	0.33	0.44	0.75
24.1 Operating cost coverage (ratio)	1.82	1.61	1.08
Production and consumption			
3.1 Water production (l/person/day)	471.00	464.00	43.00
4.1 Total water consumption (l/person/day)	187.00	186.00	158.00
4.7 Residential consumption (l/person/day)	134	134	115
Poverty and affordability			
19.1 Total revenues/service population/GNI (% GNI per capita) (average revenues)	0.91	1.02	1.02
19.2 Annual bill for households consuming 6 m^3 of water/month (US$/yr)	36.92	47.26	37.82
21.1 Ratio of industrial to residential tariff (level of cross-subsidy)	2.01	1.98	1.30

a. UNICEF and WHO 2012.

IBNET Indicator/Country: Macao SAR, China

Latest year available	2008	2009	2010
Surface area (km^2)	29	29	29
GNI per capita, Atlas method (current US$)	33,640	36,660	46,400
Total population (thousands)	518	531	544
Urban population (%)	100	100	100
Total urban population (thousands)	518	531	544
MDGs			
Access to improved water sources 2010 (%)[a]	—	—	—
Access to improved sanitation 2010 (%)[a]	—	—	—
IBNET sourced data			
Number of utilities reporting in IBNET sample	1	1	1
Population served (water), (thousands)	549	540	552
Size of the sample: Total population living in service area (water supply), (thousands)	549	540	552
Services coverage			
1.1 Water coverage (%)	100	100	100
2.1 Sewerage coverage (%)	—	—	—
Operational efficiency			
13.2 Electrical energy costs vs. operating costs (%) (share of energy cost as % of operational expenses)	11	11	10
6.1 Nonrevenue water (%)	13.00	13.00	13.00
6.2 Nonrevenue water (m^3/km/day)	56	55	54
12.3 Staff W/1,000 W population served (W/1,000 W population served)	0.40	0.50	0.50
15.1 Continuity of service (hrs/day) (duration of water supply, hours)	24.00	24.00	24.00
Financial efficiency			
8.1 Water sold that is metered (%)	100	100	100
23.1 Collection period (days)	26	34	35
23.2 Collection ratio (%)	100	101	100
18.1 Average revenue W & WW (US$/m^3 water sold)	0.57	0.58	0.58
11.1 Operational cost W & WW (US$/m^3 water sold)	0.44	0.44	0.49
24.1 Operating cost coverage (ratio)	1.28	1.33	1.17
Production and consumption			
3.1 Water production (l/person/day)	387.00	396.00	381.00
4.1 Total water consumption (l/person/day)	339.00	345.00	333.00
4.7 Residential consumption (l/person/day)	160	162	154
Poverty and affordability			
19.1 Total revenues/service population/GNI (% GNI per capita) (average revenues)	0.21	0.20	0.15
19.2 Annual bill for households consuming 6 m^3 of water/month (US$/yr)	39.91	40.32	40.23
21.1 Ratio of industrial to residential tariff (level of cross-subsidy)	1.10	1.11	1.09

a. UNICEF and WHO 2012.

IBNET Indicator/Country: Madagascar

Latest year available	2003	2004	2005
Surface area (km^2)	587,041	587,041	587,041
GNI per capita, Atlas method (current US$)	290	300	300
Total population (thousands)	16,842	17,358	17,886
Urban population (%)	28	28	29
Total urban population (thousands)	4,706	4,898	5,098
MDGs			
Access to improved water sources 2010 (%)[a]	46	46	46
Access to improved sanitation 2010 (%)[a]	15	15	15
IBNET sourced data			
Number of utilities reporting in IBNET sample	1	1	1
Population served (water), (thousands)	843	895	932
Size of the sample: Total population living in service area (water supply), (thousands)	961	986	1,032
Services coverage			
1.1 Water coverage (%)	88	91	90
2.1 Sewerage coverage (%)	20	20	20
Operational efficiency			
13.2 Electrical energy costs vs. operating costs (%) (share of energy cost as % of operational expenses)	—	—	—
6.1 Nonrevenue water (%)	36.00	33.00	34.00
6.2 Nonrevenue water (m^3/km/day)	34	29	30
12.3 Staff W/1,000 W population served (W/1,000 W population served)	—	—	—
15.1 Continuity of service (hrs/day) (duration of water supply, hours)	—	—	—
Financial efficiency			
8.1 Water sold that is metered (%)	—	—	—
23.1 Collection period (days)	—	—	—
23.2 Collection ratio (%)	—	—	—
18.1 Average revenue W & WW (US$/m^3 water sold)	—	—	—
11.1 Operational cost W & WW (US$/m^3 water sold)	—	—	—
24.1 Operating cost coverage (ratio)	—	—	—
Production and consumption			
3.1 Water production (l/person/day)	304.69	286.57	283.33
4.1 Total water consumption (l/person/day)	195.00	192.00	187.00
4.7 Residential consumption (l/person/day)	—	—	—
Poverty and affordability			
19.1 Total revenues/service population/GNI (% GNI per capita) (average revenues)	—	—	—
19.2 Annual bill for households consuming 6 m^3 of water/month (US$/yr)	—	—	—
21.1 Ratio of industrial to residential tariff (level of cross-subsidy)	—	—	—

a. UNICEF and WHO 2012.

IBNET Indicator/Country: Malawi

Latest year available	2007	2008	2009
Surface area (km^2)	118,484	118,484	118,484
GNI per capita, Atlas method (current US$)	250	290	320
Total population (thousands)	13,589	14,005	14,442
Urban population (%)	15	15	15
Total urban population (thousands)	2,038	2,101	2,166
MDGs			
Access to improved water sources 2010 (%)[a]	83	83	83
Access to improved sanitation 2010 (%)[a]	51	51	51
IBNET sourced data			
Number of utilities reporting in IBNET sample	5	5	5
Population served (water), (thousands)	1,688	1,773	1,871
Size of the sample: Total population living in service area (water supply), (thousands)	2,478	2,545	2,623
Services coverage			
1.1 Water coverage (%)	68	70	71
2.1 Sewerage coverage (%)	—	—	—
Operational efficiency			
13.2 Electrical energy costs vs. operating costs (%) (share of energy cost as % of operational expenses)	31	23	22
6.1 Nonrevenue water (%)	40.00	36.00	40.00
6.2 Nonrevenue water (m^3/km/day)	24	20	23
12.3 Staff W/1,000 W population served (W/1,000 W population served)	1.10	1.00	1.00
15.1 Continuity of service (hrs/day) (duration of water supply, hours)	19.40	19.60	21.00
Financial efficiency			
8.1 Water sold that is metered (%)	100	100	100
23.1 Collection period (days)	199	199	195
23.2 Collection ratio (%)	68	83	86
18.1 Average revenue W & WW (US$/m^3 water sold)	0.56	0.62	0.70
11.1 Operational cost W & WW (US$/m^3 water sold)	0.53	0.63	0.65
24.1 Operating cost coverage (ratio)	1.09	1.05	1.15
Production and consumption			
3.1 Water production (l/person/day)	126.00	119.00	125.00
4.1 Total water consumption (l/person/day)	76.00	76.00	75.00
4.7 Residential consumption (l/person/day)	35	37	36
Poverty and affordability			
19.1 Total revenues/service population/GNI (% GNI per capita) (average revenues)	6.21	5.93	5.99
19.2 Annual bill for households consuming 6 m^3 of water/month (US$/yr)	—	—	—
21.1 Ratio of industrial to residential tariff (level of cross-subsidy)	1.42	1.72	1.14

a. UNICEF and WHO 2012.

IBNET Indicator/Country: Malaysia

Latest year available	2007
Surface area (km^2)	330,803
GNI per capita, Atlas method (current US$)	6,550
Total population (thousands)	27,051
Urban population (%)	69
Total urban population (thousands)	18,784
MDGs	
Access to improved water sources 2010 (%)[a]	100
Access to improved sanitation 2010 (%)[a]	96
IBNET sourced data	
Number of utilities reporting in IBNET sample	7
Population served (water), (thousands)	15,154
Size of the sample: Total population living in service area (water supply), (thousands)	16,093
Services coverage	
1.1 Water coverage (%)	94
2.1 Sewerage coverage (%)	—
Operational efficiency	
13.2 Electrical energy costs vs. operating costs (%) (share of energy cost as % of operational expenses)	7
6.1 Nonrevenue water (%)	34.00
6.2 Nonrevenue water (m^3/km/day)	44
12.3 Staff W/1,000 W population served (W/1,000 W population served)	0.50
15.1 Continuity of service (hrs/day) (duration of water supply, hours)	24.00
Financial efficiency	
8.1 Water sold that is metered (%)	100
23.1 Collection period (days)	365
23.2 Collection ratio (%)	—
18.1 Average revenue W & WW (US$/m^3 water sold)	0.40
11.1 Operational cost W & WW (US$/m^3 water sold)	0.35
24.1 Operating cost coverage (ratio)	1.15
Production and consumption	
3.1 Water production (l/person/day)	535.00
4.1 Total water consumption (l/person/day)	351.00
4.7 Residential consumption (l/person/day)	227
Poverty and affordability	
19.1 Total revenues/service population/GNI (% GNI per capita) (average revenues)	0.78
19.2 Annual bill for households consuming 6 m^3 of water/month (US$/yr)	8.53
21.1 Ratio of industrial to residential tariff (level of cross-subsidy)	1.69

a. UNICEF and WHO 2012.

IBNET Indicator/Country: Mali

Latest year available	2007	2008	2009
Surface area (km²)	1,240,192	1,240,192	1,240,192
GNI per capita, Atlas method (current US$)	470	520	620
Total population (thousands)	14,021	14,460	14,910
Urban population (%)	32	33	34
Total urban population (thousands)	4,538	4,772	5,016
MDGs			
Access to improved water sources 2010 (%)[a]	64	64	64
Access to improved sanitation 2010 (%)[a]	22	22	22
IBNET sourced data			
Number of utilities reporting in IBNET sample	1	1	1
Population served (water), (thousands)	1,968	2,076	2,212
Size of the sample: Total population living in service area (water supply), (thousands)	2,815	2,958	3,107
Services coverage			
1.1 Water coverage (%)	70	70	71
2.1 Sewerage coverage (%)	—	—	—
Operational efficiency			
13.2 Electrical energy costs vs. operating costs (%) (share of energy cost as % of operational expenses)	34	23	29
6.1 Nonrevenue water (%)	26.00	26.00	27.00
6.2 Nonrevenue water (m³/km/day)	18	19	20
12.3 Staff W/1,000 W population served (W/1,000 W population served)	0.30	0.20	0.20
15.1 Continuity of service (hrs/day) (duration of water supply, hours)	24.00	24.00	24.00
Financial efficiency			
8.1 Water sold that is metered (%)	100	100	100
23.1 Collection period (days)	—	—	—
23.2 Collection ratio (%)	100	96	99
18.1 Average revenue W & WW (US$/m³ water sold)	0.64	0.75	0.65
11.1 Operational cost W & WW (US$/m³ water sold)	0.33	0.34	0.33
24.1 Operating cost coverage (ratio)	1.97	2.20	1.96
Production and consumption			
3.1 Water production (l/person/day)	98.00	99.00	99.00
4.1 Total water consumption (l/person/day)	73.00	74.00	73.00
4.7 Residential consumption (l/person/day)	66	67	66
Poverty and affordability			
19.1 Total revenues/service population/GNI (% GNI per capita) (average revenues)	3.63	3.90	2.79
19.2 Annual bill for households consuming 6 m³ of water/month (US$/yr)	—	—	—
21.1 Ratio of industrial to residential tariff (level of cross-subsidy)	1.33	1.31	1.31

a. UNICEF and WHO 2012.

IBNET Indicator/Country: Mauritania

Latest year available	2006	2007	2008
Surface area (km²)	1,025,520	1,025,520	1,025,520
GNI per capita, Atlas method (current US$)	860	960	1,080
Total population (thousands)	3,213	3,295	3,378
Urban population (%)	41	41	41
Total urban population (thousands)	1,304	1,344	1,385
MDGs			
Access to improved water sources 2010 (%)[a]	50	50	50
Access to improved sanitation 2010 (%)[a]	26	26	26
IBNET sourced data			
Number of utilities reporting in IBNET sample	1	1	1
Population served (water), (thousands)	652	551	415
Size of the sample: Total population living in service area (water supply), (thousands)	973	1,404	1,476
Services coverage			
1.1 Water coverage (%)	67	39	28
2.1 Sewerage coverage (%)	—	—	—
Operational efficiency			
13.2 Electrical energy costs vs. operating costs (%) (share of energy cost as % of operational expenses)	—	22	—
6.1 Nonrevenue water (%)	36.00	34.00	38.00
6.2 Nonrevenue water (m³/km/day)	16	18	22
12.3 Staff W/1,000 W population served (W/1,000 W population served)	1.50	2.00	2.90
15.1 Continuity of service (hrs/day) (duration of water supply, hours)	6.00	—	—
Financial efficiency			
8.1 Water sold that is metered (%)	100	100	100
23.1 Collection period (days)	321	765	—
23.2 Collection ratio (%)	116	84	—
18.1 Average revenue W & WW (US$/m³ water sold)	0.66	0.32	0.36
11.1 Operational cost W & WW (US$/m³ water sold)	0.72	0.71	—
24.1 Operating cost coverage (ratio)	0.91	0.45	—
Production and consumption			
3.1 Water production (l/person/day)	104.69	139.39	201.61
4.1 Total water consumption (l/person/day)	67.00	92.00	125.00
4.7 Residential consumption (l/person/day)	34	40	53
Poverty and affordability			
19.1 Total revenues/service population/GNI (% GNI per capita) (average revenues)	1.88	1.12	1.52
19.2 Annual bill for households consuming 6 m³ of water/month (US$/yr)	52.20	27.06	28.51
21.1 Ratio of industrial to residential tariff (level of cross-subsidy)	1.01	0.56	0.70

a. UNICEF and WHO 2012.

IBNET Indicator/Country: Mauritius

Latest year available	2004	2005	2006
Surface area (km²)	2,040	2,040	2,040
GNI per capita, Atlas method (current US$)	4,990	5,360	5,540
Total population (thousands)	1,233	1,243	1,253
Urban population (%)	40	40	41
Total urban population (thousands)	497	502	508
MDGs			
Access to improved water sources 2010 (%)[a]	99	99	99
Access to improved sanitation 2010 (%)[a]	89	89	89
IBNET sourced data			
Number of utilities reporting in IBNET sample	1	1	1
Population served (water), (thousands)	1,159	1,170	1,182
Size of the sample: Total population living in service area (water supply), (thousands)	1,159	1,171	1,182
Services coverage			
1.1 Water coverage (%)	100	100	100
2.1 Sewerage coverage (%)	—	—	—
Operational efficiency			
13.2 Electrical energy costs vs. operating costs (%) (share of energy cost as % of operational expenses)	—	—	—
6.1 Nonrevenue water (%)	53.00	52.00	54.00
6.2 Nonrevenue water (m³/km/day)	58	56	62
12.3 Staff W/1,000 W population served (W/1,000 W population served)	0.90	0.80	0.80
15.1 Continuity of service (hrs/day) (duration of water supply, hours)	24.00	24.00	24.00
Financial efficiency			
8.1 Water sold that is metered (%)	—	—	—
23.1 Collection period (days)	—	—	—
23.2 Collection ratio (%)	101	102	102
18.1 Average revenue W & WW (US$/m³ water sold)	0.34	0.33	0.32
11.1 Operational cost W & WW (US$/m³ water sold)	0.14	0.13	0.13
24.1 Operating cost coverage (ratio)	2.52	2.44	2.48
Production and consumption			
3.1 Water production (l/person/day)	436.17	441.67	460.87
4.1 Total water consumption (l/person/day)	205.00	212.00	212.00
4.7 Residential consumption (l/person/day)	160	164	164
Poverty and affordability			
19.1 Total revenues/service population/GNI (% GNI per capita) (average revenues)	0.51	0.48	0.45
19.2 Annual bill for households consuming 6 m³ of water/month (US$/yr)	—	—	—
21.1 Ratio of industrial to residential tariff (level of cross-subsidy)	2.58	2.43	2.51

a. UNICEF and WHO 2012.

IBNET Indicator/Country: Mexico

Latest year available	2007	2008	2009
Surface area (km^2)	1,964,375	1,964,375	1,964,375
GNI per capita, Atlas method (current US$)	9,070	9,630	8,500
Total population (thousands)	109,221	110,627	112,033
Urban population (%)	77	77	78
Total urban population (thousands)	83,663	84,740	87,386
MDGs			
Access to improved water sources 2010 (%)[a]	96	96	96
Access to improved sanitation 2010 (%)[a]	85	85	85
IBNET sourced data			
Number of utilities reporting in IBNET sample	13	13	13
Population served (water), (thousands)	6,130	6,147	6,321
Size of the sample: Total population living in service area (water supply), (thousands)	5,947	5,938	6,074
Services coverage			
1.1 Water coverage (%)	100	100	100
2.1 Sewerage coverage (%)	92	94	96
Operational efficiency			
13.2 Electrical energy costs vs. operating costs (%) (share of energy cost as % of operational expenses)	12	12	13
6.1 Nonrevenue water (%)	21.00	20.00	19.00
6.2 Nonrevenue water (m^3/km/day)	21	21	19
12.3 Staff W/1,000 W population served (W/1,000 W population served)	0.60	0.60	0.60
15.1 Continuity of service (hrs/day) (duration of water supply, hours)	22.80	22.90	22.90
Financial efficiency			
8.1 Water sold that is metered (%)	81	80	80
23.1 Collection period (days)	217	235	240
23.2 Collection ratio (%)	71	73	71
18.1 Average revenue W & WW (US$/m^3 water sold)	1.00	1.08	0.93
11.1 Operational cost W & WW (US$/m^3 water sold)	0.59	0.68	0.53
24.1 Operating cost coverage (ratio)	1.69	1.60	1.74
Production and consumption			
3.1 Water production (l/person/day)	263.00	268.00	258.00
4.1 Total water consumption (l/person/day)	209.00	214.00	208.00
4.7 Residential consumption (l/person/day)	171	176	173
Poverty and affordability			
19.1 Total revenues/service population/GNI (% GNI per capita) (average revenues)	0.84	0.88	0.83
19.2 Annual bill for households consuming 6 m^3 of water/month (US$/yr)	120.55	127.12	101.44
21.1 Ratio of industrial to residential tariff (level of cross-subsidy)	3.18	3.23	3.15

a. UNICEF and WHO 2012.

IBNET Indicator/Country: Micronesia

Latest year available	2011	2012	2013
Surface area (km²)	702	702	702
GNI per capita, Atlas method (current US$)	3,050	3,230	2,300
Total population (thousands)	112	112	112
Urban population (%)	23	23	23
Total urban population (thousands)	25	25	26
MDGs			
Access to improved water sources 2010 (%)[a]	94 (2000)	94 (2000)	94 (2000)
Access to improved sanitation 2010 (%)[a]	59 (2000)	59 (2000)	59 (2000)
IBNET sourced data			
Number of utilities reporting in IBNET sample	6	6	6
Population served (water), (thousands)	32	33	33
Size of the sample: Total population living in service area (water supply), (thousands)	34	35	36
Services coverage			
1.1 Water coverage (%)	38	75	79
2.1 Sewerage coverage (%)	28	35	36
Operational efficiency			
13.2 Electrical energy costs vs. operating costs (%) (share of energy cost as % of operational expenses)	—	—	—
6.1 Nonrevenue water (%)	22.00	29.00	32.00
6.2 Nonrevenue water (m³/km/day)	15	14	24
12.3 Staff W/1,000 W population served (W/1,000 W population served)	—	—	—
15.1 Continuity of service (hrs/day) (duration of water supply, hours)	20.50	24.00	15.30
Financial efficiency			
8.1 Water sold that is metered (%)	100	99	100
23.1 Collection period (days)	67	251	857
23.2 Collection ratio (%)	86	94	78
18.1 Average revenue W & WW (US$/m³ water sold)	0.58	0.69	0.55
11.1 Operational cost W & WW (US$/m³ water sold)	0.40	0.52	0.44
24.1 Operating cost coverage (ratio)	1.27	1.34	1.24
Production and consumption			
3.1 Water production (l/person/day)	740.00	325.00	398.00
4.1 Total water consumption (l/person/day)	961.00	244.00	270.00
4.7 Residential consumption (l/person/day)	—	—	—
Poverty and affordability			
19.1 Total revenues/service population/GNI (% GNI per capita) (average revenues)	6.67	1.90	2.36
19.2 Annual bill for households consuming 6 m³ of water/month (US$/yr)	498.00	—	961.60
21.1 Ratio of industrial to residential tariff (level of cross-subsidy)	—	—	—

a. UNICEF and WHO 2012.

IBNET Indicator/Country: Moldova

Latest year available	2010	2011	2012
Surface area (km^2)	33,846	33,846	33,846
GNI per capita, Atlas method (current US$)	1,820	1,980	2,070
Total population (thousands)	3,562	3,559	3,555
Urban population (%)	47	48	49
Total urban population (thousands)	1,672	1,696	1,747
MDGs			
Access to improved water sources 2010 (%)[a]	99	99	99
Access to improved sanitation 2010 (%)[a]	89	89	89
IBNET sourced data			
Number of utilities reporting in IBNET sample	39	39	39
Population served (water), (thousands)	1,171	1,182	1,184
Size of the sample: Total population living in service area (water supply), (thousands)	1,467	1,438	1,420
Services coverage			
1.1 Water coverage (%)	80	83	84
2.1 Sewerage coverage (%)	67	69	70
Operational efficiency			
13.2 Electrical energy costs vs. operating costs (%) (share of energy cost as % of operational expenses)	25	24	24
6.1 Nonrevenue water (%)	45.00	44.00	44.00
6.2 Nonrevenue water (m^3/km/day)	32	29	30
12.3 Staff W/1,000 W population served (W/1,000 W population served)	2.50	2.30	2.30
15.1 Continuity of service (hrs/day) (duration of water supply, hours)	20.30	20.80	21.30
Financial efficiency			
8.1 Water sold that is metered (%)	85	85	90
23.1 Collection period (days)	246	273	282
23.2 Collection ratio (%)	98	97	101
18.1 Average revenue W & WW (US$/m^3 water sold)	1.06	1.15	1.14
11.1 Operational cost W & WW (US$/m^3 water sold)	0.93	1.07	1.05
24.1 Operating cost coverage (ratio)	1.14	1.08	1.09
Production and consumption			
3.1 Water production (l/person/day)	248.00	236.00	240.00
4.1 Total water consumption (l/person/day)	137.00	133.00	133.00
4.7 Residential consumption (l/person/day)	106	103	103
Poverty and affordability			
19.1 Total revenues/service population/GNI (% GNI per capita) (average revenues)	2.91	2.82	2.67
19.2 Annual bill for households consuming 6 m^3 of water/month (US$/yr)	94.11	103.87	110.85
21.1 Ratio of industrial to residential tariff (level of cross-subsidy)	3.44	3.53	3.43

a. UNICEF and WHO 2012.

IBNET Indicator/Country: Mozambique

Latest year available	2005	2006	2007
Surface area (km²)	801,590	801,590	801,590
GNI per capita, Atlas method (current US$)	300	310	340
Total population (thousands)	20,770	21,291	21,811
Urban population (%)	30	30	30
Total urban population (thousands)	6,231	6,387	6,543
MDGs			
Access to improved water sources 2010 (%)[a]	47	47	47
Access to improved sanitation 2010 (%)[a]	18	18	18
IBNET sourced data			
Number of utilities reporting in IBNET sample	5	5	5
Population served (water), (thousands)	956	957	1,201
Size of the sample: Total population living in service area (water supply), (thousands)	3,166	3,199	3,211
Services coverage			
1.1 Water coverage (%)	30	30	37
2.1 Sewerage coverage (%)	—	—	—
Operational efficiency			
13.2 Electrical energy costs vs. operating costs (%) (share of energy cost as % of operational expenses)	18	17	—
6.1 Nonrevenue water (%)	58.00	56.00	59.00
6.2 Nonrevenue water (m³/km/day)	128	120	131
12.3 Staff W/1,000 W population served (W/1,000 W population served)	1.20	1.20	1.00
15.1 Continuity of service (hrs/day) (duration of water supply, hours)	14.40	16.20	19.20
Financial efficiency			
8.1 Water sold that is metered (%)	56	47	51
23.1 Collection period (days)	296	298	334
23.2 Collection ratio (%)	80	73	85
18.1 Average revenue W & WW (US$/m³ water sold)	0.55	0.57	0.69
11.1 Operational cost W & WW (US$/m³ water sold)	0.77	0.67	0.85
24.1 Operating cost coverage (ratio)	0.72	0.85	0.82
Production and consumption			
3.1 Water production (l/person/day)	245.24	240.91	212.20
4.1 Total water consumption (l/person/day)	103.00	106.00	87.00
4.7 Residential consumption (l/person/day)	—	—	—
Poverty and affordability			
19.1 Total revenues/service population/GNI (% GNI per capita) (average revenues)	6.89	7.11	6.44
19.2 Annual bill for households consuming 6 m³ of water/month (US$/yr)	—	—	—
21.1 Ratio of industrial to residential tariff (level of cross-subsidy)	—	—	—

a. UNICEF and WHO 2012.

IBNET Indicator/Country: Namibia

Latest year available	2007	2008	2009
Surface area (km^2)	824,116	824,116	824,116
GNI per capita, Atlas method (current US$)	3,970	4,080	4,190
Total population (thousands)	2,159	2,200	2,242
Urban population (%)	36	37	37
Total urban population (thousands)	781	808	835
MDGs			
Access to improved water sources 2010 (%)[a]	93	93	93
Access to improved sanitation 2010 (%)[a]	32	32	32
IBNET sourced data			
Number of utilities reporting in IBNET sample	1	1	1
Population served (water), (thousands)	57	60	63
Size of the sample: Total population living in service area (water supply), (thousands)	57	60	63
Services coverage			
1.1 Water coverage (%)	100	100	100
2.1 Sewerage coverage (%)	100	100	100
Operational efficiency			
13.2 Electrical energy costs vs. operating costs (%) (share of energy cost as % of operational expenses)	1	0	1
6.1 Nonrevenue water (%)	15.00	17.00	14.00
6.2 Nonrevenue water (m^3/km/day)	5	6	5
12.3 Staff W/1,000 W population served (W/1,000 W population served)	0.50	0.50	0.50
15.1 Continuity of service (hrs/day) (duration of water supply, hours)	24.00	24.00	24.00
Financial efficiency			
8.1 Water sold that is metered (%)	100	100	100
23.1 Collection period (days)	0	0	0
23.2 Collection ratio (%)	71	69	69
18.1 Average revenue W & WW (US$/m^3 water sold)	2.23	2.05	2.20
11.1 Operational cost W & WW (US$/m^3 water sold)	1.64	1.44	1.49
24.1 Operating cost coverage (ratio)	1.35	1.42	1.48
Production and consumption			
3.1 Water production (l/person/day)	210.00	212.00	208.00
4.1 Total water consumption (l/person/day)	178.00	177.00	179.00
4.7 Residential consumption (l/person/day)	80	80	80
Poverty and affordability			
19.1 Total revenues/service population/GNI (% GNI per capita) (average revenues)	3.65	3.25	3.43
19.2 Annual bill for households consuming 6 m^3 of water/month (US$/yr)	108.94	101.69	109.09
21.1 Ratio of industrial to residential tariff (level of cross-subsidy)	—	—	—

a. UNICEF and WHO 2012.

IBNET Indicator/Country: Netherlands Antilles

Latest year available	2005	2006
Surface area (km^2)	37,354	37,354
GNI per capita, Atlas method (current US$)	33,000	33,000
Total population (thousands)	186	189
Urban population (%)	92	92
Total urban population (thousands)	171	174
MDGs		
Access to improved water sources 2010 (%)[a]	100	100
Access to improved sanitation 2010 (%)[a]	100	100
IBNET sourced data		
Number of utilities reporting in IBNET sample	1	1
Population served (water), (thousands)	133	134
Size of the sample: Total population living in service area (water supply), (thousands)	133	134
Services coverage		
1.1 Water coverage (%)	100	100
2.1 Sewerage coverage (%)	—	—
Operational efficiency		
13.2 Electrical energy costs vs. operating costs (%) (share of energy cost as % of operational expenses)	—	—
6.1 Nonrevenue water (%)	30.00	29.00
6.2 Nonrevenue water (m^3/km/day)	4	4
12.3 Staff W/1,000 W population served (W/1,000 W population served)	2.70	2.60
15.1 Continuity of service (hrs/day) (duration of water supply, hours)	24.00	24.00
Financial efficiency		
8.1 Water sold that is metered (%)	100	100
23.1 Collection period (days)	126	100
23.2 Collection ratio (%)	—	—
18.1 Average revenue W & WW (US$/m^3 water sold)	7.60	8.10
11.1 Operational cost W & WW (US$/m^3 water sold)	5.09	6.44
24.1 Operating cost coverage (ratio)	1.49	1.26
Production and consumption		
3.1 Water production (l/person/day)	255.71	252.11
4.1 Total water consumption (l/person/day)	179.00	179.00
4.7 Residential consumption (l/person/day)	132	128
Poverty and affordability		
19.1 Total revenues/service population/GNI (% GNI per capita) (average revenues)	1.50	1.60
19.2 Annual bill for households consuming 6 m^3 of water/month (US$/yr)	69.72	72.74
21.1 Ratio of industrial to residential tariff (level of cross-subsidy)	1.06	1.02

a. UNICEF and WHO 2012.

IBNET Indicator/Country: New Zealand

Latest year available	2005	2006	2007
Surface area (km^2)	270,467	270,467	270,467
GNI per capita, Atlas method (current US$)	24,990	25,570	27,320
Total population (thousands)	4,134	4,185	4,228
Urban population (%)	86	86	86
Total urban population (thousands)	3,563	3,612	3,655
MDGs			
Access to improved water sources 2010 (%)[a]	100	100	100
Access to improved sanitation 2010 (%)[a]	100	100	100
IBNET sourced data			
Number of utilities reporting in IBNET sample	1	1	1
Population served (water), (thousands)	419	425	431
Size of the sample: Total population living in service area (water supply), (thousands)	419	425	431
Services coverage			
1.1 Water coverage (%)	100	100	100
2.1 Sewerage coverage (%)	100	102	100
Operational efficiency			
13.2 Electrical energy costs vs. operating costs (%) (share of energy cost as % of operational expenses)	—	—	—
6.1 Nonrevenue water (%)	18.00	12.00	12.00
6.2 Nonrevenue water (m^3/km/day)	12	1	1
12.3 Staff W/1,000 W population served (W/1,000 W population served)	—	—	—
15.1 Continuity of service (hrs/day) (duration of water supply, hours)	24.00	24.00	24.00
Financial efficiency			
8.1 Water sold that is metered (%)	100	100	100
23.1 Collection period (days)	—	—	—
23.2 Collection ratio (%)	—	—	—
18.1 Average revenue W & WW (US$/m^3 water sold)	1.89	1.36	1.90
11.1 Operational cost W & WW (US$/m^3 water sold)	1.89	1.36	1.90
24.1 Operating cost coverage (ratio)	1.00	1.00	1.00
Production and consumption			
3.1 Water production (l/person/day)	358.54	396.59	388.64
4.1 Total water consumption (l/person/day)	294.00	349.00	342.00
4.7 Residential consumption (l/person/day)	189	164	168
Poverty and affordability			
19.1 Total revenues/service population/GNI (% GNI per capita) (average revenues)	0.81	0.68	0.87
19.2 Annual bill for households consuming 6 m^3 of water/month (US$/yr)	—	—	—
21.1 Ratio of industrial to residential tariff (level of cross-subsidy)	—	—	—

a. UNICEF and WHO 2012.

IBNET Indicator/Country: Nicaragua

Latest year available	2003	2004	2005
Surface area (km^2)	130,373	130,373	130,373
GNI per capita, Atlas method (current US$)	1,000	1,090	1,170
Total population (thousands)	5,288	5,356	5,424
Urban population (%)	55	55	56
Total urban population (thousands)	2,931	2,946	3,032
MDGs			
Access to improved water sources 2010 (%)[a]	85	85	85
Access to improved sanitation 2010 (%)[a]	52	52	52
IBNET sourced data			
Number of utilities reporting in IBNET sample	1	1	1
Population served (water), (thousands)	2,916	2,998	2,969
Size of the sample: Total population living in service area (water supply), (thousands)	3,190	3,153	3,153
Services coverage			
1.1 Water coverage (%)	91	95	94
2.1 Sewerage coverage (%)	35	35	34
Operational efficiency			
13.2 Electrical energy costs vs. operating costs (%) (share of energy cost as % of operational expenses)	—	—	40
6.1 Nonrevenue water (%)	57.00	60.00	58.00
6.2 Nonrevenue water (m^3/km/day)	89	—	—
12.3 Staff W/1,000 W population served (W/1,000 W population served)	—	—	—
15.1 Continuity of service (hrs/day) (duration of water supply, hours)	20.00	20.00	20.00
Financial efficiency			
8.1 Water sold that is metered (%)	—	—	69
23.1 Collection period (days)	—	—	151
23.2 Collection ratio (%)	—	—	82
18.1 Average revenue W & WW (US$/m^3 water sold)	1.41	—	0.42
11.1 Operational cost W & WW (US$/m^3 water sold)	—	—	0.38
24.1 Operating cost coverage (ratio)	—	—	1.11
Production and consumption			
3.1 Water production (l/person/day)	239.53	250.00	259.52
4.1 Total water consumption (l/person/day)	103.00	100.00	109.00
4.7 Residential consumption (l/person/day)	60	60	60
Poverty and affordability			
19.1 Total revenues/service population/GNI (% GNI per capita) (average revenues)	5.30	—	1.43
19.2 Annual bill for households consuming 6 m^3 of water/month (US$/yr)	—	—	—
21.1 Ratio of industrial to residential tariff (level of cross-subsidy)	—	—	—

a. UNICEF and WHO 2012.

IBNET Indicator/Country: Niger

Latest year available	2007	2008	2009
Surface area (km²)	1,267,000	1,267,000	1,267,000
GNI per capita, Atlas method (current US$)	290	330	340
Total population (thousands)	13,946	14,450	14,972
Urban population (%)	17	17	17
Total urban population (thousands)	2,383	2,494	2,611
MDGs			
Access to improved water sources 2010 (%)[a]	49	49	49
Access to improved sanitation 2010 (%)[a]	9	9	9
IBNET sourced data			
Number of utilities reporting in IBNET sample	1	1	1
Population served (water), (thousands)	1,665	1,736	1,819
Size of the sample: Total population living in service area (water supply), (thousands)	2,337	2,418	2,509
Services coverage			
1.1 Water coverage (%)	71	72	72
2.1 Sewerage coverage (%)	—	—	—
Operational efficiency			
13.2 Electrical energy costs vs. operating costs (%) (share of energy cost as % of operational expenses)	2	2	2
6.1 Nonrevenue water (%)	17.00	15.00	16.00
6.2 Nonrevenue water (m³/km/day)	8	7	7
12.3 Staff W/1,000 W population served (W/1,000 W population served)	0.40	0.30	0.40
15.1 Continuity of service (hrs/day) (duration of water supply, hours)	22.00	22.00	22.00
Financial efficiency			
8.1 Water sold that is metered (%)	100	100	100
23.1 Collection period (days)	227	243	273
23.2 Collection ratio (%)	90	92	87
18.1 Average revenue W & WW (US$/m³ water sold)	0.55	0.64	0.58
11.1 Operational cost W & WW (US$/m³ water sold)	0.43	0.48	0.47
24.1 Operating cost coverage (ratio)	1.28	1.33	1.22
Production and consumption			
3.1 Water production (l/person/day)	77.00	77.00	79.00
4.1 Total water consumption (l/person/day)	64.00	65.00	67.00
4.7 Residential consumption (l/person/day)	51.00	51.00	52.00
Poverty and affordability			
19.1 Total revenues/service population/GNI (% GNI per capita) (average revenues)	4.43	4.60	4.17
19.2 Annual bill for households consuming 6 m³ of water/month (US$/yr)	31.69	37.10	33.13
21.1 Ratio of industrial to residential tariff (level of cross-subsidy)	6.48	6.61	6.79

a. UNICEF and WHO 2012.

IBNET Indicator/Country: Pakistan

Latest year available	2010	2011	2012
Surface area (km²)	803,940	803,940	803,940
GNI per capita, Atlas method (current US$)	1,060	1,140	1,260
Total population (thousands)	167,442	170,494	173,593
Urban population (%)	35	36	36
Total urban population (thousands)	57,868	61,378	62,494
MDGs			
Access to improved water sources 2010 (%)[a]	92	92	92
Access to improved sanitation 2010 (%)[a]	48	48	48
IBNET sourced data			
Number of utilities reporting in IBNET sample	5	1	1
Population served (water), (thousands)	24,438	217	217
Size of the sample: Total population living in service area (water supply), (thousands)	30,620	1,244	1,284
Services coverage			
1.1 Water coverage (%)	80	17	17
2.1 Sewerage coverage (%)	78	52	51
Operational efficiency			
13.2 Electrical energy costs vs. operating costs (%) (share of energy cost as % of operational expenses)	42	43	39
6.1 Nonrevenue water (%)	36.00	57.00	57.00
6.2 Nonrevenue water (m³/km/day)	183	—	316
12.3 Staff W/1,000 W population served (W/1,000 W population served)	0.60	1.90	2.50
15.1 Continuity of service (hrs/day) (duration of water supply, hours)	9.90	10.80	—
Financial efficiency			
8.1 Water sold that is metered (%)	15	—	—
23.1 Collection period (days)	1,168	166	159
23.2 Collection ratio (%)	53	55	62
18.1 Average revenue W & WW (US$/m³ water sold)	0.10	0.01	0.02
11.1 Operational cost W & WW (US$/m³ water sold)	0.11	0.19	0.31
24.1 Operating cost coverage (ratio)	0.89	0.06	0.08
Production and consumption			
3.1 Water production (l/person/day)	233.00	240.00	289.00
4.1 Total water consumption (l/person/day)	153.00	124.00	124.00
4.7 Residential consumption (l/person/day)	127	113	113
Poverty and affordability			
19.1 Total revenues/service population/GNI (% GNI per capita) (average revenues)	0.53	0.04	0.07
19.2 Annual bill for households consuming 6 m³ of water/month (US$/yr)	4.28	3.41	2.81
21.1 Ratio of industrial to residential tariff (level of cross-subsidy)	30.47	34.22	0.41

a. UNICEF and WHO 2012.

IBNET Indicator/Country: Panama

Latest year available	2004	2005	2006
Surface area (km²)	75,517	75,517	75,517
GNI per capita, Atlas method (current US$)	4,300	4,640	4,940
Total population (thousands)	3,182	3,238	3,295
Urban population (%)	70	71	72
Total urban population (thousands)	2,221	2,293	2,359
MDGs			
Access to improved water sources 2010 (%)[a]	97	97	97
Access to improved sanitation 2010 (%)[a]	74 (2000)	74 (2000)	74 (2000)
IBNET sourced data			
Number of utilities reporting in IBNET sample	1	1	1
Population served (water), (thousands)	2,037	1,889	2,381
Size of the sample: Total population living in service area (water supply), (thousands)	2,243	2,303	2,372
Services coverage			
1.1 Water coverage (%)	91	82	100
2.1 Sewerage coverage (%)	52	45	48
Operational efficiency			
13.2 Electrical energy costs vs. operating costs (%) (share of energy cost as % of operational expenses)	60	38	44
6.1 Nonrevenue water (%)	44.00	43.00	39.00
6.2 Nonrevenue water (m³/km/day)	120	112	100
12.3 Staff W/1,000 W population served (W/1,000 W population served)	—	0.80	0.60
15.1 Continuity of service (hrs/day) (duration of water supply, hours)	20.00	22.00	
Financial efficiency			
8.1 Water sold that is metered (%)	44	46	43
23.1 Collection period (days)	342	152	112
23.2 Collection ratio (%)	—	78	111
18.1 Average revenue W & WW (US$/m³ water sold)	0.27	0.26	0.25
11.1 Operational cost W & WW (US$/m³ water sold)	0.11	0.19	0.18
24.1 Operating cost coverage (ratio)	2.38	1.39	1.44
Production and consumption			
3.1 Water production (l/person/day)	619.64	684.21	549.18
4.1 Total water consumption (l/person/day)	347.00	390.00	335.00
4.7 Residential consumption (l/person/day)	258	151	112
Poverty and affordability			
19.1 Total revenues/service population/GNI (% GNI per capita) (average revenues)	0.80	0.80	0.62
19.2 Annual bill for households consuming 6 m³ of water/month (US$/yr)	—	—	—
21.1 Ratio of industrial to residential tariff (level of cross-subsidy)	—	0.35	—

a. UNICEF and WHO 2012.

IBNET Indicator/Country: Papua New Guniea

Latest year available	2011	2012	2013
Surface area (km²)	462,840	462,840	462,840
GNI per capita, Atlas method (current US$)	1,480	1,790	1,800
Total population (thousands)	7,014	7,169	7,324
Urban population (%)	13	13	13
Total urban population (thousands)	877	896	916
MDGs			
Access to improved water sources 2010 (%)[a]	40	40	40
Access to improved sanitation 2010 (%)[a]	45	45	45
IBNET sourced data			
Number of utilities reporting in IBNET sample	1	2	2
Population served (water), (thousands)	500	770	840
Size of the sample: Total population living in service area (water supply), (thousands)	415	739	731
Services coverage			
1.1 Water coverage (%)	83	96	87
2.1 Sewerage coverage (%)	40	21	18
Operational efficiency			
13.2 Electrical energy costs vs. operating costs (%) (share of energy cost as % of operational expenses)	—	—	—
6.1 Nonrevenue water (%)	51.00	51.00	51.00
6.2 Nonrevenue water (m³/km/day)	51	52	45
12.3 Staff W/1,000 W population served (W/1,000 W population served)	—	—	—
15.1 Continuity of service (hrs/day) (duration of water supply, hours)	24.00	24.00	24.00
Financial efficiency			
8.1 Water sold that is metered (%)	100	100	93
23.1 Collection period (days)	144	152	143
23.2 Collection ratio (%)	101	83	93
18.1 Average revenue W & WW (US$/m³ water sold)	1.85	1.85	2.03
11.1 Operational cost W & WW (US$/m³ water sold)	1.00	1.33	1.38
24.1 Operating cost coverage (ratio)	1.85	1.39	1.47
Production and consumption			
3.1 Water production (l/person/day)	321.00	303.00	309.00
4.1 Total water consumption (l/person/day)	156.00	150.00	151.00
4.7 Residential consumption (l/person/day)	—	—	—
Poverty and affordability			
19.1 Total revenues/service population/GNI (% GNI per capita) (average revenues)	7.12	5.66	6.22
19.2 Annual bill for households consuming 6 m³ of water/month (US$/yr)	727.72	—	847.27
21.1 Ratio of industrial to residential tariff (level of cross-subsidy)	—	—	—

a. UNICEF and WHO 2012.

IBNET Indicator/Country: Paraguay

Latest year available	2003	2004	2005
Surface area (km²)	406,752	406,752	406,752
GNI per capita, Atlas method (current US$)	1,010	1,070	1,260
Total population (thousands)	5,676	5,787	5,898
Urban population (%)	57	58	58
Total urban population (thousands)	3,248	3,348	3,421
MDGs			
Access to improved water sources 2010 (%)[a]	86	86	86
Access to improved sanitation 2010 (%)[a]	71	71	71
IBNET sourced data			
Number of utilities reporting in IBNET sample	1	4	4
Population served (water), (thousands)	903	770	717
Size of the sample: Total population living in service area (water supply), (thousands)	1,188	1,421	1,002
Services coverage			
1.1 Water coverage (%)	76	54	72
2.1 Sewerage coverage (%)	43	39	32
Operational efficiency			
13.2 Electrical energy costs vs. operating costs (%) (share of energy cost as % of operational expenses)	19	19	17
6.1 Nonrevenue water (%)	52.00	45.00	44.00
6.2 Nonrevenue water (m³/km/day)	45	44	42
12.3 Staff W/1,000 W population served (W/1,000 W population served)	—	—	0.50
15.1 Continuity of service (hrs/day) (duration of water supply, hours)	24.00	24.00	24.00
Financial efficiency			
8.1 Water sold that is metered (%)	—	90	91
23.1 Collection period (days)	54	173	170
23.2 Collection ratio (%)	—	—	—
18.1 Average revenue W & WW (US$/m³ water sold)	0.86	0.37	0.36
11.1 Operational cost W & WW (US$/m³ water sold)	0.16	0.17	0.17
24.1 Operating cost coverage (ratio)	5.34	2.23	2.15
Production and consumption			
3.1 Water production (l/person/day)	270.83	440.00	433.93
4.1 Total water consumption (l/person/day)	130.00	242.00	243.00
4.7 Residential consumption (l/person/day)	121	206	205
Poverty and affordability			
19.1 Total revenues/service population/GNI (% GNI per capita) (average revenues)	4.04	3.05	2.53
19.2 Annual bill for households consuming 6 m³ of water/month (US$/yr)	—	—	—
21.1 Ratio of industrial to residential tariff (level of cross-subsidy)	—	—	0.10

a. UNICEF and WHO 2012.

IBNET Indicator/Country: Peru

Latest year available	2006	2007	2008
Surface area (km^2)	1,285,216	1,285,216	1,285,216
GNI per capita, Atlas method (current US$)	3,130	3,380	4,020
Total population (thousands)	27,866	28,166	28,463
Urban population (%)	75	76	76
Total urban population (thousands)	21,014	21,346	21,678
MDGs			
Access to improved water sources 2010 (%)[a]	85	85	85
Access to improved sanitation 2010 (%)[a]	71	71	71
IBNET sourced data			
Number of utilities reporting in IBNET sample	49	49	49
Population served (water), (thousands)	13,955	14,599	14,990
Size of the sample: Total population living in service area (water supply), (thousands)	16,689	17,076	17,443
Services coverage			
1.1 Water coverage (%)	84	85	86
2.1 Sewerage coverage (%)	76	77	78
Operational efficiency			
13.2 Electrical energy costs vs. operating costs (%) (share of energy cost as % of operational expenses)	23	23	23
6.1 Nonrevenue water (%)	43.00	42.00	42.00
6.2 Nonrevenue water (m^3/km/day)	63	61	62
12.3 Staff W/1,000 W population served (W/1,000 W population served)	—	—	—
15.1 Continuity of service (hrs/day) (duration of water supply, hours)	16.50	15.10	15.11
Financial efficiency			
8.1 Water sold that is metered (%)	60	63	63
23.1 Collection period (days)	128	114	67
23.2 Collection ratio (%)	—	—	101
18.1 Average revenue W & WW (US$/m^3 water sold)	0.53	0.57	0.68
11.1 Operational cost W & WW (US$/m^3 water sold)	0.49	0.50	0.58
24.1 Operating cost coverage (ratio)	1.08	1.14	1.17
Production and consumption			
3.1 Water production (l/person/day)	245.61	234.48	232.76
4.1 Total water consumption (l/person/day)	140.00	136.00	135.00
4.7 Residential consumption (l/person/day)	76	—	—
Poverty and affordability			
19.1 Total revenues/service population/GNI (% GNI per capita) (average revenues)	0.87	0.84	0.83
19.2 Annual bill for households consuming 6 m^3 of water/month (US$/yr)	—	—	—
21.1 Ratio of industrial to residential tariff (level of cross-subsidy)	1.00	1.00	1.00

a. UNICEF and WHO 2012.

IBNET Indicator/Country: Philippines

Latest year available	2007	2008	2009
Surface area (km^2)	300,000	300,000	300,000
GNI per capita, Atlas method (current US$)	1,510	1,770	1,870
Total population (thousands)	88,653	90,173	91,703
Urban population (%)	48	48	49
Total urban population (thousands)	42,801	43,646	44,499
MDGs			
Access to improved water sources 2010 (%)[a]	92	92	92
Access to improved sanitation 2010 (%)[a]	74	74	74
IBNET sourced data			
Number of utilities reporting in IBNET sample	4	25	25
Population served (water), (thousands)	13,740	16,084	15,886
Size of the sample: Total population living in service area (water supply), (thousands)	16,815	20,831	20,522
Services coverage			
1.1 Water coverage (%)	82	77	77
2.1 Sewerage coverage (%)	7	6	6
Operational efficiency			
13.2 Electrical energy costs vs. operating costs (%) (share of energy cost as % of operational expenses)	20	18	23
6.1 Nonrevenue water (%)	49.00	45.00	43.00
6.2 Nonrevenue water (m^3/km/day)	284	207	194
12.3 Staff W/1,000 W population served (W/1,000 W population served)	0.40	0.40	0.40
15.1 Continuity of service (hrs/day) (duration of water supply, hours)	20.70	22.80	21.70
Financial efficiency			
8.1 Water sold that is metered (%)	100	100	100
23.1 Collection period (days)	36	47	40
23.2 Collection ratio (%)	98	98	99
18.1 Average revenue W & WW (US$/m^3 water sold)	0.48	0.53	0.54
11.1 Operational cost W & WW (US$/m^3 water sold)	0.23	0.24	0.22
24.1 Operating cost coverage (ratio)	2.12	0.15	2.40
Production and consumption			
3.1 Water production (l/person/day)	294.00	281.00	271.00
4.1 Total water consumption (l/person/day)	151.00	154.00	156.00
4.7 Residential consumption (l/person/day)	111	118	117
Poverty and affordability			
19.1 Total revenues/service population/GNI (% GNI per capita) (average revenues)	1.75	1.68	1.64
19.2 Annual bill for households consuming 6 m^3 of water/month (US$/yr)	26.37	37.79	36.22
21.1 Ratio of industrial to residential tariff (level of cross-subsidy)	1.74	1.77	2.48

a. UNICEF and WHO 2012.

IBNET Indicator/Country: Poland

Latest year available	2008	2009	2010
Surface area (km^2)	312,685	312,685	312,685
GNI per capita, Atlas method (current US$)	11,870	12,190	12,400
Total population (thousands)	38,126	38,152	38,184
Urban population (%)	62	61	61
Total urban population (thousands)	23,447	23,440	23,437
MDGs			
Access to improved water sources 2010 (%)[a]	100	100	100
Access to improved sanitation 2010 (%)[a]	99	99	99
IBNET sourced data			
Number of utilities reporting in IBNET sample	31	31	31
Population served (water), (thousands)	8,533	8,550	8,624
Size of the sample: Total population living in service area (water supply), (thousands)	8,927	8,919	8,981
Services coverage			
1.1 Water coverage (%)	96	96	96
2.1 Sewerage coverage (%)	89	90	90
Operational efficiency			
13.2 Electrical energy costs vs. operating costs (%) (share of energy cost as % of operational expenses)	9	10	10
6.1 Nonrevenue water (%)	17.00	16.00	15.00
6.2 Nonrevenue water (m^3/km/day)	10	9	8
12.3 Staff W/1,000 W population served (W/1,000 W population served)	0.60	0.60	0.60
15.1 Continuity of service (hrs/day) (duration of water supply, hours)	24.00	24.00	24.00
Financial efficiency			
8.1 Water sold that is metered (%)	100	100	99
23.1 Collection period (days)	158	71	74
23.2 Collection ratio (%)	111	111	112
18.1 Average revenue W & WW (US$/m^3 water sold)	2.60	1.93	1.92
11.1 Operational cost W & WW (US$/m^3 water sold)	1.92	1.40	1.41
24.1 Operating cost coverage (ratio)	1.35	1.38	1.36
Production and consumption			
3.1 Water production (l/person/day)	190.00	183.00	179.00
4.1 Total water consumption (l/person/day)	158.00	154.00	152.00
4.7 Residential consumption (l/person/day)	117	114	113
Poverty and affordability			
19.1 Total revenues/service population/GNI (% GNI per capita) (average revenues)	1.26	0.89	0.86
19.2 Annual bill for households consuming 6 m^3 of water/month (US$/yr)	99.43	71.80	70.59
21.1 Ratio of industrial to residential tariff (level of cross-subsidy)	1.80	1.79	1.78

a. UNICEF and WHO 2012.

IBNET Indicator/Country: Romania

Latest year available	2008	2009	2010
Surface area (km²)	238,391	238,391	238,391
GNI per capita, Atlas method (current US$)	8,290	8,250	8,010
Total population (thousands)	21,514	21,480	21,438
Urban population (%)	53	53	53
Total urban population (thousands)	11,402	11,385	11,362
MDGs			
Access to improved water sources 2010 (%)[a]	99	99	99
Access to improved sanitation 2010 (%)[a]	88 (2000)	88 (2000)	88 (2000)
IBNET sourced data			
Number of utilities reporting in IBNET sample	19	20	20
Population served (water), (thousands)	3,968	4,518	4,952
Size of the sample: Total population living in service area (water supply), (thousands)	4,715	5,478	5,926
Services coverage			
1.1 Water coverage (%)	84	82	84
2.1 Sewerage coverage (%)	69	65	62
Operational efficiency			
13.2 Electrical energy costs vs. operating costs (%) (share of energy cost as % of operational expenses)	15	12	11
6.1 Nonrevenue water (%)	49.00	49.00	51.00
6.2 Nonrevenue water (m³/km/day)	61	48	43
12.3 Staff W/1,000 W population served (W/1,000 W population served)	2.00	2.10	2.00
15.1 Continuity of service (hrs/day) (duration of water supply, hours)	24.00	24.00	24.00
Financial efficiency			
8.1 Water sold that is metered (%)	92	92	95
23.1 Collection period (days)	80	80	87
23.2 Collection ratio (%)	107	108	112
18.1 Average revenue W & WW (US$/m³ water sold)	1.20	1.08	1.02
11.1 Operational cost W & WW (US$/m³ water sold)	1.12	0.98	0.94
24.1 Operating cost coverage (ratio)	1.07	1.10	1.08
Production and consumption			
3.1 Water production (l/person/day)	371.00	336.00	311.00
4.1 Total water consumption (l/person/day)	191.00	172.00	153.00
4.7 Residential consumption (l/person/day)	111	106	103
Poverty and affordability			
19.1 Total revenues/service population/GNI (% GNI per capita) (average revenues)	1.01	0.82	0.71
19.2 Annual bill for households consuming 6 m³ of water/month (US$/yr)	129.34	98.51	82.27
21.1 Ratio of industrial to residential tariff (level of cross-subsidy)	1.37	1.40	1.51

a. UNICEF and WHO 2012.

IBNET Indicator/Country: Russian Federation

Latest year available	2009	2010	2011	2012
Surface area (km^2)	17,098,242	17,098,242	17,098,242	17,098,242
GNI per capita, Atlas method (current US$)	9,290	10,000	10,810	12,700
Total population (thousands)	141,910	142,389	142,960	143,297
Urban population (%)	74	74	74	75
Total urban population (thousands)	105,013	105,368	105,790	107,473
MDGs				
Access to improved water sources 2010 (%)[a]	97	97	97	97
Access to improved sanitation 2010 (%)[a]	70	70	70	70
IBNET sourced data				
Number of utilities reporting in IBNET sample	82	95	80	88
Population served (water), (thousands)	43,164	45,306	54,837	55,098
Size of the sample: Total population living in service area (water supply), (thousands)	42,070	45,249	54,837	55,098
Services coverage				
1.1 Water coverage (%)	100	100	100	100
2.1 Sewerage coverage (%)	94	94	95	94
Operational efficiency				
13.2 Electrical energy costs vs. operating costs (%) (share of energy cost as % of operational expenses)	18	25	22	19
6.1 Nonrevenue water (%)	21.00	24.00	23.00	24.00
6.2 Nonrevenue water (m^3/km/day)	56	60	54	52
12.3 Staff W/1,000 W population served (W/1,000 W population served)	1.40	1.30	1.30	1.30
15.1 Continuity of service (hrs/day) (duration of water supply, hours)	24.00	24.00	24.00	24.00
Financial efficiency				
8.1 Water sold that is metered (%)	—	74	—	—
23.1 Collection period (days)	92	95	99	111
23.2 Collection ratio (%)	91	92	88	91
18.1 Average revenue W & WW (US$/m^3 water sold)	0.63	0.73	0.85	0.93
11.1 Operational cost W & WW (US$/m^3 water sold)	0.45	0.52	0.60	0.66
24.1 Operating cost coverage (ratio)	1.40	1.40	1.41	1.38
Production and consumption				
3.1 Water production (l/person/day)	436.00	415.00	384.00	375.00
4.1 Total water consumption (l/person/day)	343.00	316.00	296.00	284.00
4.7 Residential consumption (l/person/day)	216	199	182	172
Poverty and affordability				
19.1 Total revenues/service population/GNI (% GNI per capita) (average revenues)	0.85	0.84	0.85	0.76
19.2 Annual bill for households consuming 6 m^3 of water/month (US$/yr)	27.83	32.17	37.64	38.86
21.1 Ratio of industrial to residential tariff (level of cross-subsidy)	1.45	1.27	1.24	1.22

a. UNICEF and WHO 2012.

IBNET Indicator/Country: Rwanda

Latest year available	2003	2004	2005
Surface area (km^2)	26,338	26,338	26,338
GNI per capita, Atlas method (current US$)	210	230	270
Total population (thousands)	8,858	9,010	9,202
Urban population (%)	16	17	18
Total urban population (thousands)	1,419	1,510	1,610
MDGs			
Access to improved water sources 2010 (%)[a]	65	65	65
Access to improved sanitation 2010 (%)[a]	55	55	55
IBNET sourced data			
Number of utilities reporting in IBNET sample	1	1	1
Population served (water), (thousands)	2,085	2,232	2,394
Size of the sample: Total population living in service area (water supply), (thousands)	1,843	1,973	2,010
Services coverage			
1.1 Water coverage (%)	113	113	119
2.1 Sewerage coverage (%)	—	—	—
Operational efficiency			
13.2 Electrical energy costs vs. operating costs (%) (share of energy cost as % of operational expenses)	42	31	46
6.1 Nonrevenue water (%)	51.00	44.00	38.00
6.2 Nonrevenue water (m^3/km/day)	12	9	7
12.3 Staff W/1,000 W population served (W/1,000 W population served)	0.60	0.60	0.60
15.1 Continuity of service (hrs/day) (duration of water supply, hours)	12.00	12.00	12.00
Financial efficiency			
8.1 Water sold that is metered (%)	100	100	100
23.1 Collection period (days)	—	50	438
23.2 Collection ratio (%)	100	100	144
18.1 Average revenue W & WW (US$/m^3 water sold)	0.62	0.57	0.42
11.1 Operational cost W & WW (US$/m^3 water sold)	0.19	0.34	0.51
24.1 Operating cost coverage (ratio)	3.36	1.65	0.82
Production and consumption			
3.1 Water production (l/person/day)	24.49	19.64	17.74
4.1 Total water consumption (l/person/day)	12.00	11.00	11.00
4.7 Residential consumption (l/person/day)	—	—	—
Poverty and affordability			
19.1 Total revenues/service population/GNI (% GNI per capita) (average revenues)	1.29	1.00	0.62
19.2 Annual bill for households consuming 6 m^3 of water/month (US$/yr)	33.71	31.54	32.60
21.1 Ratio of industrial to residential tariff (level of cross-subsidy)	—	—	—

a. UNICEF and WHO 2012.

IBNET Indicator/Country: Senegal

Latest year available	2007	2008	2009
Surface area (km^2)	196,722	196,722	196,722
GNI per capita, Atlas method (current US$)	900	1,020	1,030
Total population (thousands)	11,475	11,787	12,107
Urban population (%)	42	42	42
Total urban population (thousands)	4,819	4,903	5,068
MDGs			
Access to improved water sources 2010 (%)[a]	72	72	72
Access to improved sanitation 2010 (%)[a]	52	52	52
IBNET sourced data			
Number of utilities reporting in IBNET sample	1	1	1
Population served (water), (thousands)	5,124	5,358	5,504
Size of the sample: Total population living in service area (water supply), (thousands)	5,445	5,848	6,281
Services coverage			
1.1 Water coverage (%)	94	92	88
2.1 Sewerage coverage (%)	—	—	—
Operational efficiency			
13.2 Electrical energy costs vs. operating costs (%) (share of energy cost as % of operational expenses)	46	52	50
6.1 Nonrevenue water (%)	20.00	21.00	21.00
6.2 Nonrevenue water (m^3/km/day)	9	10	10
12.3 Staff W/1,000 W population served (W/1,000 W population served)	0.20	0.20	0.20
15.1 Continuity of service (hrs/day) (duration of water supply, hours)	24.00	24.00	24.00
Financial efficiency			
8.1 Water sold that is metered (%)	100	100	100
23.1 Collection period (days)	18	14	13
23.2 Collection ratio (%)	90	91	94
18.1 Average revenue W & WW (US$/m^3 water sold)	1.15	1.40	1.25
11.1 Operational cost W & WW (US$/m^3 water sold)	0.44	0.55	0.51
24.1 Operating cost coverage (ratio)	2.51	2.55	2.40
Production and consumption			
3.1 Water production (l/person/day)	72.00	71.00	69.00
4.1 Total water consumption (l/person/day)	58.00	56.00	55.00
4.7 Residential consumption (l/person/day)	49	47	47
Poverty and affordability			
19.1 Total revenues/service population/GNI (% GNI per capita) (average revenues)	2.71	2.81	2.44
19.2 Annual bill for households consuming 6 m^3 of water/month (US$/yr)	—	—	—
21.1 Ratio of industrial to residential tariff (level of cross-subsidy)	4.33	4.33	4.33

a. UNICEF and WHO 2012.

IBNET Indicator/Country: The Seychelles

Latest year available	2004	2005	2006
Surface area (km²)	8,240	9,820	11,150
GNI per capita, Atlas method (current US$)	8,190	9,000	9,000
Total population (thousands)	83	83	85
Urban population (%)	51	52	52
Total urban population (thousands)	42	43	44
MDGs			
Access to improved water sources 2010 (%)[a]	100	100	100
Access to improved sanitation 2010 (%)[a]	98	98	98
IBNET sourced data			
Number of utilities reporting in IBNET sample	1	1	1
Population served (water), (thousands)	79	80	80
Size of the sample: Total population living in service area (water supply), (thousands)	79	80	80
Services coverage			
1.1 Water coverage (%)	100	99	100
2.1 Sewerage coverage (%)	15	15	20
Operational efficiency			
13.2 Electrical energy costs vs. operating costs (%) (share of energy cost as % of operational expenses)	—	—	—
6.1 Nonrevenue water (%)	17.00	20.00	14.00
6.2 Nonrevenue water (m³/km/day)	18	25	15
12.3 Staff W/1,000 W population served (W/1,000 W population served)	5.30	5.20	5.20
15.1 Continuity of service (hrs/day) (duration of water supply, hours)	24.00	24.00	24.00
Financial efficiency			
8.1 Water sold that is metered (%)	50	45	45
23.1 Collection period (days)	—	—	—
23.2 Collection ratio (%)	99	100	100
18.1 Average revenue W & WW (US$/m³ water sold)	1.06	0.77	0.79
11.1 Operational cost W & WW (US$/m³ water sold)	2.04	1.69	1.77
24.1 Operating cost coverage (ratio)	0.52	0.45	0.44
Production and consumption			
3.1 Water production (l/person/day)	418.07	473.75	438.37
4.1 Total water consumption (l/person/day)	347.00	379.00	377.00
4.7 Residential consumption (l/person/day)	—	—	—
Poverty and affordability			
19.1 Total revenues/service population/GNI (% GNI per capita) (average revenues)	1.64	1.18	1.21
19.2 Annual bill for households consuming 6 m³ of water/month (US$/yr)	86.60	81.96	83.00
21.1 Ratio of industrial to residential tariff (level of cross-subsidy)	—	—	—

a. UNICEF and WHO 2012.

IBNET Indicator/Country: Singapore

Latest year available	2007	2008
Surface area (km²)	705	705
GNI per capita, Atlas method (current US$)	33,760	35,750
Total population (thousands)	4,589	4,839
Urban population (%)	100	100
Total urban population (thousands)	4,589	4,839
MDGs		
Access to improved water sources 2010 (%)[a]	100	100
Access to improved sanitation 2010 (%)[a]	100	100
IBNET sourced data		
Number of utilities reporting in IBNET sample	1	1
Population served (water), (thousands)	4,589	4,840
Size of the sample: Total population living in service area (water supply), (thousands)	4,589	4,840
Services coverage		
1.1 Water coverage (%)	100	100
2.1 Sewerage coverage (%)	100	100
Operational efficiency		
13.2 Electrical energy costs vs. operating costs (%) (share of energy cost as % of operational expenses)	—	—
6.1 Nonrevenue water (%)	4.00	4.00
6.2 Nonrevenue water (m³/km/day)	10	9
12.3 Staff W/1,000 W population served (W/1,000 W population served)	0.30	0.30
15.1 Continuity of service (hrs/day) (duration of water supply, hours)	24.00	24.00
Financial efficiency		
8.1 Water sold that is metered (%)	100	100
23.1 Collection period (days)	—	—
23.2 Collection ratio (%)	—	—
18.1 Average revenue W & WW (US$/m³ water sold)	—	—
11.1 Operational cost W & WW (US$/m³ water sold)	—	—
24.1 Operating cost coverage (ratio)	—	—
Production and consumption		
3.1 Water production (l/person/day)	283.33	272.92
4.1 Total water consumption (l/person/day)	272.00	262.00
4.7 Residential consumption (l/person/day)	158	154
Poverty and affordability		
19.1 Total revenues/service population/GNI (% GNI per capita) (average revenues)	—	—
19.2 Annual bill for households consuming 6 m³ of water/month (US$/yr)	90.54	101.74
21.1 Ratio of industrial to residential tariff (level of cross-subsidy)	—	—

a. UNICEF and WHO 2012.

IBNET Indicator/Country: Slovak Republic

Latest year available	2005	2006	2007
Surface area (km²)	49,035	49,035	49,035
GNI per capita, Atlas method (current US$)	11,040	12,550	14,410
Total population (thousands)	5,387	5,391	5,397
Urban population (%)	56	55	55
Total urban population (thousands)	3,027	2,965	2,969
MDGs			
Access to improved water sources 2010 (%)[a]	100	100	100
Access to improved sanitation 2010 (%)[a]	100	100	100
IBNET sourced data			
Number of utilities reporting in IBNET sample	5	6	7
Population served (water), (thousands)	2,732	3,430	3,664
Size of the sample: Total population living in service area (water supply), (thousands)	3,479	4,209	4,533
Services coverage			
1.1 Water coverage (%)	79	81	81
2.1 Sewerage coverage (%)	53	57	56
Operational efficiency			
13.2 Electrical energy costs vs. operating costs (%) (share of energy cost as % of operational expenses)	—	—	—
6.1 Nonrevenue water (%)	11.00	18.00	18.00
6.2 Nonrevenue water (m³/km/day)	14	16	14
12.3 Staff W/1,000 W population served (W/1,000 W population served)	—	0.50	0.50
15.1 Continuity of service (hrs/day) (duration of water supply, hours)	24.00	24.00	24.00
Financial efficiency			
8.1 Water sold that is metered (%)	100	100	100
23.1 Collection period (days)	72	70	80
23.2 Collection ratio (%)	104	103	102
18.1 Average revenue W & WW (US$/m³ water sold)	1.34	1.50	1.81
11.1 Operational cost W & WW (US$/m³ water sold)	0.94	1.04	1.28
24.1 Operating cost coverage (ratio)	1.43	1.44	1.42
Production and consumption			
3.1 Water production (l/person/day)	161.80	184.15	181.71
4.1 Total water consumption (l/person/day)	144.00	151.00	149.00
4.7 Residential consumption (l/person/day)	98	101	102
Poverty and affordability			
19.1 Total revenues/service population/GNI (% GNI per capita) (average revenues)	0.64	0.66	0.68
19.2 Annual bill for households consuming 6 m³ of water/month (US$/yr)	91.13	112.34	133.96
21.1 Ratio of industrial to residential tariff (level of cross-subsidy)	1.32	0.98	0.98

a. UNICEF and WHO 2012.

IBNET Indicator/Country: South Africa

Latest year available	2007	2008	2009
Surface area (km^2)	1,221,037	1,221,037	1,221,037
GNI per capita, Atlas method (current US$)	5,760	5,850	5,730
Total population (thousands)	48,257	48,793	49,320
Urban population (%)	60	61	61
Total urban population (thousands)	29,037	29,583	30,129
MDGs			
Access to improved water sources 2010 (%)[a]	91	91	91
Access to improved sanitation 2010 (%)[a]	79	79	79
IBNET sourced data			
Number of utilities reporting in IBNET sample	4	4	4
Population served (water), (thousands)	7,887	8,074	8,169
Size of the sample: Total population living in service area (water supply), (thousands)	10,102	10,301	10,499
Services coverage			
1.1 Water coverage (%)	78	78	78
2.1 Sewerage coverage (%)	49	50	53
Operational efficiency			
13.2 Electrical energy costs vs. operating costs (%) (share of energy cost as % of operational expenses)	1	2	2
6.1 Nonrevenue water (%)	32.00	37.00	37.00
6.2 Nonrevenue water (m^3/km/day)	27	35	35
12.3 Staff W/1,000 W population served (W/1,000 W population served)	0.30	0.10	0.30
15.1 Continuity of service (hrs/day) (duration of water supply, hours)	24.00	24.00	24.00
Financial efficiency			
8.1 Water sold that is metered (%)	99	100	100
23.1 Collection period (days)	314	368	284
23.2 Collection ratio (%)	97	98	100
18.1 Average revenue W & WW (US$/m^3 water sold)	1.15	1.21	1.26
11.1 Operational cost W & WW (US$/m^3 water sold)	1.28	1.27	1.41
24.1 Operating cost coverage (ratio)	0.99	0.95	0.89
Production and consumption			
3.1 Water production (l/person/day)	306.00	384.00	386.00
4.1 Total water consumption (l/person/day)	207.00	244.00	242.00
4.7 Residential consumption (l/person/day)	132	191	190
Poverty and affordability			
19.1 Total revenues/service population/GNI (% GNI per capita) (average revenues)	1.51	1.84	1.94
19.2 Annual bill for households consuming 6 m^3 of water/month (US$/yr)	—	—	—
21.1 Ratio of industrial to residential tariff (level of cross-subsidy)	1.12	1.08	0.96

a. UNICEF and WHO 2012.

IBNET Indicator/Country: Sri Lanka

Latest year available	2007	2008	2009
Surface area (km^2)	65,610	65,610	65,610
GNI per capita, Atlas method (current US$)	1,540	1,770	1,970
Total population (thousands)	20,039	20,217	20,450
Urban population (%)	15	15	15
Total urban population (thousands)	3,026	3,053	3,088
MDGs			
Access to improved water sources 2010 (%)[a]	91	91	91
Access to improved sanitation 2010 (%)[a]	92	92	92
IBNET sourced data			
Number of utilities reporting in IBNET sample	1	1	1
Population served (water), (thousands)	3,300	3,510	3,680
Size of the sample: Total population living in service area (water supply), (thousands)	4,400	4,500	4,600
Services coverage			
1.1 Water coverage (%)	75	78	80
2.1 Sewerage coverage (%)	2	3	3
Operational efficiency			
13.2 Electrical energy costs vs. operating costs (%) (share of energy cost as % of operational expenses)	21	—	—
6.1 Nonrevenue water (%)	33.00	32.00	31.00
6.2 Nonrevenue water (m^3/km/day)	38	37	37
12.3 Staff W/1,000 W population served (W/1,000 W population served)	2.60	2.50	2.40
15.1 Continuity of service (hrs/day) (duration of water supply, hours)	20.00	21.00	22.00
Financial efficiency			
8.1 Water sold that is metered (%)	—	—	—
23.1 Collection period (days)	—	—	—
23.2 Collection ratio (%)	100	99	94
18.1 Average revenue W & WW (US$/m^3 water sold)	0.24	0.25	0.32
11.1 Operational cost W & WW (US$/m^3 water sold)	0.23	0.28	0.35
24.1 Operating cost coverage (ratio)	1.03	0.89	0.90
Production and consumption			
3.1 Water production (l/person/day)	352.00	343.00	334.00
4.1 Total water consumption (l/person/day)	236.00	235.00	231.00
4.7 Residential consumption (l/person/day)	—	132	131
Poverty and affordability			
19.1 Total revenues/service population/GNI (% GNI per capita) (average revenues)	1.34	1.21	1.37
19.2 Annual bill for households consuming 6 m^3 of water/month (US$/yr)	—	—	—
21.1 Ratio of industrial to residential tariff (level of cross-subsidy)	15.38	12.93	—

a. UNICEF and WHO 2012.

IBNET Indicator/Country: Sudan

Latest year available	2007	2008	2009
Surface area (km²)	2,505,813	2,506,000	2,506,000
GNI per capita, Atlas method (current US$)	550	900	1,120
Total population (thousands)	31,935	32,438	32,971
Urban population (%)	33	33	33
Total urban population (thousands)	10,538	10,705	10,880
MDGs			
Access to improved water sources 2010 (%)[a]	58	58	58
Access to improved sanitation 2010 (%)[a]	26	26	26
IBNET sourced data			
Number of utilities reporting in IBNET sample	5	5	5
Population served (water), (thousands)	3,501	3,820	4,241
Size of the sample: Total population living in service area (water supply), (thousands)	6,480	6,723	7,173
Services coverage			
1.1 Water coverage (%)	54	57	59
2.1 Sewerage coverage (%)	—	—	—
Operational efficiency			
13.2 Electrical energy costs vs. operating costs (%) (share of energy cost as % of operational expenses)	25	36	24
6.1 Nonrevenue water (%)	15.00	7.00	9.00
6.2 Nonrevenue water (m³/km/day)	25	11	14
12.3 Staff W/1,000 W population served (W/1,000 W population served)	1.20	1.00	0.90
15.1 Continuity of service (hrs/day) (duration of water supply, hours)	18.60	18.20	18.20
Financial efficiency			
8.1 Water sold that is metered (%)	32	32	39
23.1 Collection period (days)	171	171	136
23.2 Collection ratio (%)	—	—	—
18.1 Average revenue W & WW (US$/m³ water sold)	0.46	0.41	0.36
11.1 Operational cost W & WW (US$/m³ water sold)	0.23	0.21	0.18
24.1 Operating cost coverage (ratio)	2.00	1.98	2.03
Production and consumption			
3.1 Water production (l/person/day)	231.00	227.00	264.00
4.1 Total water consumption (l/person/day)	197.00	211.00	241.00
4.7 Residential consumption (l/person/day)	121	131	151
Poverty and affordability			
19.1 Total revenues/service population/GNI (% GNI per capita) (average revenues)	6.01	3.51	2.83
19.2 Annual bill for households consuming 6 m³ of water/month (US$/yr)	122.45	118.81	169.23
21.1 Ratio of industrial to residential tariff (level of cross-subsidy)	8.38	10.01	12.98

a. UNICEF and WHO 2012.

IBNET Indicator/Country: Swaziland

Latest year available	2007	2008	2009
Surface area (km²)	17,364	17,364	17,364
GNI per capita, Atlas method (current US$)	3,030	3,100	2,670
Total population (thousands)	1,020	1,032	1,044
Urban population (%)	22	22	21
Total urban population (thousands)	221	222	224
MDGs			
Access to improved water sources 2010 (%)[a]	71	71	71
Access to improved sanitation 2010 (%)[a]	57	57	57
IBNET sourced data			
Number of utilities reporting in IBNET sample	1	1	1
Population served (water), (thousands)	270	285	285
Size of the sample: Total population living in service area (water supply), (thousands)	290	300	300
Services coverage			
1.1 Water coverage (%)	93	95	95
2.1 Sewerage coverage (%)	33	38	38
Operational efficiency			
13.2 Electrical energy costs vs. operating costs (%) (share of energy cost as % of operational expenses)	8	10	14
6.1 Nonrevenue water (%)	39.00	37.00	40.00
6.2 Nonrevenue water (m³/km/day)	28	26	30
12.3 Staff W/1,000 W population served (W/1,000 W population served)	—	—	—
15.1 Continuity of service (hrs/day) (duration of water supply, hours)	24.00	24.00	24.00
Financial efficiency			
8.1 Water sold that is metered (%)	100	100	100
23.1 Collection period (days)	86	62	65
23.2 Collection ratio (%)	96	99	97
18.1 Average revenue W & WW (US$/m³ water sold)	1.53	1.40	1.56
11.1 Operational cost W & WW (US$/m³ water sold)	1.66	1.33	1.48
24.1 Operating cost coverage (ratio)	0.92	1.06	1.06
Production and consumption			
3.1 Water production (l/person/day)	183.00	183.00	192.00
4.1 Total water consumption (l/person/day)	112.00	115.00	115.00
4.7 Residential consumption (l/person/day)	—	77	77
Poverty and affordability			
19.1 Total revenues/service population/GNI (% GNI per capita) (average revenues)	2.06	1.90	2.45
19.2 Annual bill for households consuming 6 m³ of water/month (US$/yr)	79.10	74.75	79.37
21.1 Ratio of industrial to residential tariff (level of cross-subsidy)	—	3.02	3.00

a. UNICEF and WHO 2012.

IBNET Indicator/Country: Tajikistan

Latest year available	2003	2004	2005
Surface area (km²)	143,100	143,100	143,100
GNI per capita, Atlas method (current US$)	210	280	340
Total population (thousands)	6,337	6,391	6,453
Urban population (%)	26	26	26
Total urban population (thousands)	1,675	1,689	1,704
MDGs			
Access to improved water sources 2010 (%)[a]	64	64	64
Access to improved sanitation 2010 (%)[a]	94	94	94
IBNET sourced data			
Number of utilities reporting in IBNET sample	9	9	9
Population served (water), (thousands)	1,029	1,042	1,090
Size of the sample: Total population living in service area (water supply), (thousands)	1,112	1,128	1,179
Services coverage			
1.1 Water coverage (%)	93	92	92
2.1 Sewerage coverage (%)	60	60	59
Operational efficiency			
13.2 Electrical energy costs vs. operating costs (%) (share of energy cost as % of operational expenses)	—	—	—
6.1 Nonrevenue water (%)	35.00	35.00	36.00
6.2 Nonrevenue water (m³/km/day)	199	208	226
12.3 Staff W/1,000 W population served (W/1,000 W population served)	1.50	1.40	1.20
15.1 Continuity of service (hrs/day) (duration of water supply, hours)	22.40	21.73	21.29
Financial efficiency			
8.1 Water sold that is metered (%)	1	2	1
23.1 Collection period (days)	326	263	273
23.2 Collection ratio (%)	52	47	42
18.1 Average revenue W & WW (US$/m³ water sold)	0.01	0.02	0.03
11.1 Operational cost W & WW (US$/m³ water sold)	0.01	0.01	0.02
24.1 Operating cost coverage (ratio)	1.39	1.40	1.42
Production and consumption			
3.1 Water production (l/person/day)	796.92	833.85	865.63
4.1 Total water consumption (l/person/day)	518.00	542.00	554.00
4.7 Residential consumption (l/person/day)	313	328	336
Poverty and affordability			
19.1 Total revenues/service population/GNI (% GNI per capita) (average revenues)	0.90	1.41	1.78
19.2 Annual bill for households consuming 6 m³ of water/month (US$/yr)	0.87	1.66	1.62
21.1 Ratio of industrial to residential tariff (level of cross-subsidy)	20.09	22.21	23.56

a. UNICEF and WHO 2012.

IBNET Indicator/Country: Tanzania

Latest year available	2007	2008	2009
Surface area (km^2)	945,087	945,087	945,087
GNI per capita, Atlas method (current US$)	410	460	500
Total population (thousands)	41,068	42,268	43,525
Urban population (%)	25	25	26
Total urban population (thousands)	10,300	10,567	11,108
MDGs			
Access to improved water sources 2010 (%)[a]	—	—	—
Access to improved sanitation 2010 (%)[a]	—	—	—
IBNET sourced data			
Number of utilities reporting in IBNET sample	20	20	20
Population served (water), (thousands)	4,749	5,675	4,749
Size of the sample: Total population living in service area (water supply), (thousands)	5,977	6,959	5,977
Services coverage			
1.1 Water coverage (%)	79	82	83
2.1 Sewerage coverage (%)	5	4	4
Operational efficiency			
13.2 Electrical energy costs vs. operating costs (%) (share of energy cost as % of operational expenses)	—	—	—
6.1 Nonrevenue water (%)	45.00	36.00	46.00
6.2 Nonrevenue water (m^3/km/day)	52	35	44
12.3 Staff W/1,000 W population served (W/1,000 W population served)	0.50	0.50	0.30
15.1 Continuity of service (hrs/day) (duration of water supply, hours)	15.00	17.60	17.50
Financial efficiency			
8.1 Water sold that is metered (%)	100	100	—
23.1 Collection period (days)	81	95	263
23.2 Collection ratio (%)	88	97	103
18.1 Average revenue W & WW (US$/m^3 water sold)	0.35	0.24	0.39
11.1 Operational cost W & WW (US$/m^3 water sold)	0.40	0.29	0.44
24.1 Operating cost coverage (ratio)	0.88	0.83	0.87
Production and consumption			
3.1 Water production (l/person/day)	116.00	95.00	82.00
4.1 Total water consumption (l/person/day)	64.00	61.00	44.00
4.7 Residential consumption (l/person/day)	43	—	29
Poverty and affordability			
19.1 Total revenues/service population/GNI (% GNI per capita) (average revenues)	1.99	1.16	1.25
19.2 Annual bill for households consuming 6 m^3 of water/month (US$/yr)	—	—	—
21.1 Ratio of industrial to residential tariff (level of cross-subsidy)	1.02	1.01	—

a. UNICEF and WHO 2012.

IBNET Indicator/Country: Togo

Latest year available	2007	2008	2009
Surface area (km²)	56,785	56,785	56,785
GNI per capita, Atlas method (current US$)	420	470	450
Total population (thousands)	5,653	5,777	5,902
Urban population (%)	36	37	37
Total urban population (thousands)	2,042	2,114	2,187
MDGs			
Access to improved water sources 2010 (%)[a]	61	61	61
Access to improved sanitation 2010 (%)[a]	13	13	13
IBNET sourced data			
Number of utilities reporting in IBNET sample	1	1	1
Population served (water), (thousands)	1,340	1,419	1,536
Size of the sample: Total population living in service area (water supply), (thousands)	2,497	2,600	2,800
Services coverage			
1.1 Water coverage (%)	54	55	55
2.1 Sewerage coverage (%)	8	8	7
Operational efficiency			
13.2 Electrical energy costs vs. operating costs (%) (share of energy cost as % of operational expenses)	6	7	7
6.1 Nonrevenue water (%)	15.00	16.00	15.00
6.2 Nonrevenue water (m³/km/day)	2	2	2
12.3 Staff W/1,000 W population served (W/1,000 W population served)	0.50	0.50	0.40
15.1 Continuity of service (hrs/day) (duration of water supply, hours)	24.00	24.00	24.00
Financial efficiency			
8.1 Water sold that is metered (%)	100	100	100
23.1 Collection period (days)	86	90	112
23.2 Collection ratio (%)	88	98	91
18.1 Average revenue W & WW (US$/m³ water sold)	0.70	0.81	0.71
11.1 Operational cost W & WW (US$/m³ water sold)	1.60	2.02	1.57
24.1 Operating cost coverage (ratio)	0.44	0.40	0.45
Production and consumption			
3.1 Water production (l/person/day)	36.00	36.00	36.00
4.1 Total water consumption (l/person/day)	31.00	30.00	32.00
4.7 Residential consumption (l/person/day)	26	26	27
Poverty and affordability			
19.1 Total revenues/service population/GNI (% GNI per capita) (average revenues)	1.89	1.89	1.84
19.2 Annual bill for households consuming 6 m³ of water/month (US$/yr)	—	—	—
21.1 Ratio of industrial to residential tariff (level of cross-subsidy)	—	—	—

a. UNICEF and WHO 2012.

IBNET Indicator/Country: Tunisia

Latest year available	2008	2009	2010
Surface area (km^2)	163,610	163,610	163,610
GNI per capita, Atlas method (current US$)	3,900	4,100	4,150
Total population (thousands)	10,329	10,440	10,549
Urban population (%)	66	66	66
Total urban population (thousands)	6,787	6,880	6,973
MDGs			
Access to improved water sources 2010 (%)[a]	—	—	—
Access to improved sanitation 2010 (%)[a]	—	—	—
IBNET sourced data			
Number of utilities reporting in IBNET sample	2	2	2
Population served (water), (thousands)	8,527	8,647	8,780
Size of the sample: Total population living in service area (water supply), (thousands)	10,380	10,489	10,615
Services coverage			
1.1 Water coverage (%)	82	82	83
2.1 Sewerage coverage (%)	—	—	—
Operational efficiency			
13.2 Electrical energy costs vs. operating costs (%) (share of energy cost as % of operational expenses)	10	11	12
6.1 Nonrevenue water (%)	25.00	25.00	26.00
6.2 Nonrevenue water (m^3/km/day)	7	8	8
12.3 Staff W/1,000 W population served (W/1,000 W population served)	0.80	0.80	0.80
15.1 Continuity of service (hrs/day) (duration of water supply, hours)	24.00	24.00	24.00
Financial efficiency			
8.1 Water sold that is metered (%)	101	101	101
23.1 Collection period (days)	—	195	183
23.2 Collection ratio (%)	—	—	—
18.1 Average revenue W & WW (US$/m^3 water sold)	—	0.44	0.40
11.1 Operational cost W & WW (US$/m^3 water sold)	—	0.59	0.50
24.1 Operating cost coverage (ratio)	0.82	0.76	0.81
Production and consumption			
3.1 Water production (l/person/day)	153.00	156.00	164.00
4.1 Total water consumption (l/person/day)	115.00	116.00	121.00
4.7 Residential consumption (l/person/day)	86	86	90
Poverty and affordability			
19.1 Total revenues/service population/GNI (% GNI per capita) (average revenues)	—	0.45	0.43
19.2 Annual bill for households consuming 6 m^3 of water/month (US$/yr)	17.23	15.00	13.58
21.1 Ratio of industrial to residential tariff (level of cross-subsidy)	1.00	1.00	1.00

a. UNICEF and WHO 2012.

IBNET Indicator/Country: Turkey

Latest year available	2006	2007	2008
Surface area (km²)	783,562	783,562	783,562
GNI per capita, Atlas method (current US$)	7,470	8,440	9,260
Total population (thousands)	69,064	69,993	70,924
Urban population (%)	68	68	69
Total urban population (thousands)	46,797	47,749	48,710
MDGs			
Access to improved water sources 2010 (%)[a]	100	100	100
Access to improved sanitation 2010 (%)[a]	90	90	90
IBNET sourced data			
Number of utilities reporting in IBNET sample	17	20	10
Population served (water), (thousands)	2,398	2,619	905
Size of the sample: Total population living in service area (water supply), (thousands)	2,328	2,641	990
Services coverage			
1.1 Water coverage (%)	100	99	100
2.1 Sewerage coverage (%)	94	95	94
Operational efficiency			
13.2 Electrical energy costs vs. operating costs (%) (share of energy cost as % of operational expenses)	54	51	41
6.1 Nonrevenue water (%)	56.00	62.00	59.00
6.2 Nonrevenue water (m³/km/day)	44	63	43
12.3 Staff W/1,000 W population served (W/1,000 W population served)	—	0.70	0.50
15.1 Continuity of service (hrs/day) (duration of water supply, hours)	24.00	24.00	24.00
Financial efficiency			
8.1 Water sold that is metered (%)	97	96	100
23.1 Collection period (days)	—	139	108
23.2 Collection ratio (%)	96	94	90
18.1 Average revenue W & WW (US$/m³ water sold)	1.04	1.36	1.21
11.1 Operational cost W & WW (US$/m³ water sold)	0.72	1.09	0.93
24.1 Operating cost coverage (ratio)	1.37	1.25	1.25
Production and consumption			
3.1 Water production (l/person/day)	204.55	242.11	278.05
4.1 Total water consumption (l/person/day)	90.00	92.00	114.00
4.7 Residential consumption (l/person/day)	72	73	87
Poverty and affordability			
19.1 Total revenues/service population/GNI (% GNI per capita) (average revenues)	0.46	0.54	0.54
19.2 Annual bill for households consuming 6 m³ of water/month (US$/yr)	66.05	79.77	85.00
21.1 Ratio of industrial to residential tariff (level of cross-subsidy)	2.31	2.25	1.72

a. UNICEF and WHO 2012.

IBNET Indicator/Country: Uganda

Latest year available	2007	2008	2009
Surface area (km²)	241,038	241,038	241,038
GNI per capita, Atlas method (current US$)	380	420	400
Total population (thousands)	30,340	31,339	32,368
Urban population (%)	14	14	15
Total urban population (thousands)	4,251	4,511	4,783
MDGs			
Access to improved water sources 2010 (%)[a]	72	72	72
Access to improved sanitation 2010 (%)[a]	34	34	34
IBNET sourced data			
Number of utilities reporting in IBNET sample	1	1	1
Population served (water), (thousands)	1,803	1,944	2,137
Size of the sample: Total population living in service area (water supply), (thousands)	2,540	2,700	2,946
Services coverage			
1.1 Water coverage (%)	71	72	73
2.1 Sewerage coverage (%)	7	6	6
Operational efficiency			
13.2 Electrical energy costs vs. operating costs (%) (share of energy cost as % of operational expenses)	—	18	18
6.1 Nonrevenue water (%)	33.00	34.00	36.00
6.2 Nonrevenue water (m³/km/day)	17	18	14
12.3 Staff W/1,000 W population served (W/1,000 W population served)	0.70	0.60	0.60
15.1 Continuity of service (hrs/day) (duration of water supply, hours)	23.00	23.00	23.00
Financial efficiency			
8.1 Water sold that is metered (%)	—	—	—
23.1 Collection period (days)	518	525	449
23.2 Collection ratio (%)	92	92	99
18.1 Average revenue W & WW (US$/m³ water sold)	1.05	1.29	1.10
11.1 Operational cost W & WW (US$/m³ water sold)	0.78	1.04	0.82
24.1 Operating cost coverage (ratio)	1.34	1.24	1.34
Production and consumption			
3.1 Water production (l/person/day)	92.54	90.91	89.06
4.1 Total water consumption (l/person/day)	62.00	60.00	57.00
4.7 Residential consumption (l/person/day)	33	31	30
Poverty and affordability			
19.1 Total revenues/service population/GNI (% GNI per capita) (average revenues)	6.25	6.73	5.72
19.2 Annual bill for households consuming 6 m³ of water/month (US$/yr)	—	—	—
21.1 Ratio of industrial to residential tariff (level of cross-subsidy)	1.00	1.00	1.00

a. UNICEF and WHO 2012.

IBNET Indicator/Country: Ukraine

Latest year available	2005	2006	2007
Surface area (km^2)	603,500	603,500	603,500
GNI per capita, Atlas method (current US$)	1,540	1,950	2,570
Total population (thousands)	47,105	46,788	46,509
Urban population (%)	68	68	68
Total urban population (thousands)	31,937	31,750	31,589
MDGs			
Access to improved water sources 2010 (%)[a]	98	98	98
Access to improved sanitation 2010 (%)[a]	94	94	94
IBNET sourced data			
Number of utilities reporting in IBNET sample	16	16	16
Population served (water), (thousands)	2,703	2,721	2,736
Size of the sample: Total population living in service area (water supply), (thousands)	3,452	3,432	3,411
Services coverage			
1.1 Water coverage (%)	78	79	80
2.1 Sewerage coverage (%)	63	64	67
Operational efficiency			
13.2 Electrical energy costs vs. operating costs (%) (share of energy cost as % of operational expenses)	30	33	36
6.1 Nonrevenue water (%)	45.00	44.00	45.00
6.2 Nonrevenue water (m^3/km/day)	77	77	75
12.3 Staff W/1,000 W population served (W/1,000 W population served)	2.20	2.10	2.10
15.1 Continuity of service (hrs/day) (duration of water supply, hours)	22.00	22.00	22.00
Financial efficiency			
8.1 Water sold that is metered (%)	27	31	36
23.1 Collection period (days)	278	251	225
23.2 Collection ratio (%)	92	84	92
18.1 Average revenue W & WW (US$/m^3 water sold)	0.25	0.32	0.44
11.1 Operational cost W & WW (US$/m^3 water sold)	0.30	0.37	0.48
24.1 Operating cost coverage (ratio)	0.84	0.87	0.91
Production and consumption			
3.1 Water production (l/person/day)	496.36	494.64	480.00
4.1 Total water consumption (l/person/day)	273.00	277.00	264.00
4.7 Residential consumption (l/person/day)	231	224	208
Poverty and affordability			
19.1 Total revenues/service population/GNI (% GNI per capita) (average revenues)	1.62	1.66	1.65
19.2 Annual bill for households consuming 6 m^3 of water/month (US$/yr)	18.07	29.49	37.32
21.1 Ratio of industrial to residential tariff (level of cross-subsidy)	5.30	4.93	3.25

a. UNICEF and WHO 2012.

IBNET Indicator/Country: United States

Latest year available	2011
Surface area (km^2)	9,827,000
GNI per capita, Atlas method (current US$)	48,040
Total population (thousands)	311,592
Urban population (%)	82
Total urban population (thousands)	255,505
MDGs	
Access to improved water sources 2010 (%)[a]	99
Access to improved sanitation 2010 (%)[a]	100
IBNET sourced data	
Number of utilities reporting in IBNET sample	1
Population served (water), (thousands)	311
Size of the sample: Total population living in service area (water supply), (thousands)	311
Services coverage	
1.1 Water coverage (%)	100
2.1 Sewerage coverage (%)	46
Operational efficiency	
13.2 Electrical energy costs vs. operating costs (%) (share of energy cost as % of operational expenses)	7
6.1 Nonrevenue water (%)	13.00
6.2 Nonrevenue water (m^3/km/day)	10
12.3 Staff W/1,000 W population served (W/1,000 W population served)	0.70
15.1 Continuity of service (hrs/day) (duration of water supply, hours)	24.00
Financial efficiency	
8.1 Water sold that is metered (%)	100
23.1 Collection period (days)	415
23.2 Collection ratio (%)	168
18.1 Average revenue W & WW (US$/m^3 water sold)	1.36
11.1 Operational cost W & WW (US$/m^3 water sold)	0.92
24.1 Operating cost coverage (ratio)	1.48
Production and consumption	
3.1 Water production (l/person/day)	699.00
4.1 Total water consumption (l/person/day)	610.00
4.7 Residential consumption (l/person/day)	218
Poverty and affordability	
19.1 Total revenues/service population/GNI (% GNI per capita) (average revenues)	0.63
19.2 Annual bill for households consuming 6 m^3 of water/month (US$/yr)	1464.00
21.1 Ratio of industrial to residential tariff (level of cross-subsidy)	0.49

a. UNICEF and WHO 2012.

IBNET Indicator/Country: Uruguay

Latest year available	2009	2010	2011
Surface area (km^2)	176,215	176,215	176,215
GNI per capita, Atlas method (current US$)	8,520	10,110	11,700
Total population (thousands)	3,324	3,334	3,345
Urban population (%)	92	92	92
Total urban population (thousands)	3,049	3,063	3,081
MDGs			
Access to improved water sources 2010 (%)[a]	100	100	100
Access to improved sanitation 2010 (%)[a]	100	100	100
IBNET sourced data			
Number of utilities reporting in IBNET sample	1	1	1
Population served (water), (thousands)	3,014	3,027	3,033
Size of the sample: Total population living in service area (water supply), (thousands)	3,261	3,271	3,282
Services coverage			
1.1 Water coverage (%)	98	98	90
2.1 Sewerage coverage (%)	23	24	23
Operational efficiency			
13.2 Electrical energy costs vs. operating costs (%) (share of energy cost as % of operational expenses)	13	12	14
6.1 Nonrevenue water (%)	53.00	49.00	49.00
6.2 Nonrevenue water (m^3/km/day)	34	32	31
12.3 Staff W/1,000 W population served (W/1,000 W population served)	—	—	—
15.1 Continuity of service (hrs/day) (duration of water supply, hours)	24.00	24.00	24.00
Financial efficiency			
8.1 Water sold that is metered (%)	96	97	97
23.1 Collection period (days)	54	61	76
23.2 Collection ratio (%)	103	11	107
18.1 Average revenue W & WW (US$/m^3 water sold)	1.84	1.80	1.94
11.1 Operational cost W & WW (US$/m^3 water sold)	1.40	1.48	1.48
24.1 Operating cost coverage (ratio)	1.32	1.22	1.31
Production and consumption			
3.1 Water production (l/person/day)	277.00	280.00	311.00
4.1 Total water consumption (l/person/day)	130.00	142.00	157.00
4.7 Residential consumption (l/person/day)	104	114	128
Poverty and affordability			
19.1 Total revenues/service population/GNI (% GNI per capita) (average revenues)	1.02	0.92	0.95
19.2 Annual bill for households consuming 6 m^3 of water/month (US$/yr)	—	124.26	127.36
21.1 Ratio of industrial to residential tariff (level of cross-subsidy)	2.21	2.83	3.13

a. UNICEF and WHO 2012.

IBNET Indicator/Country: Uzbekistan

Latest year available	2008	2009	2010
Surface area (km²)	447,400	447,400	447,400
GNI per capita, Atlas method (current US$)	960	1,130	1,300
Total population (thousands)	26,167	26,486	26,868
Urban population (%)	36	36	36
Total urban population (thousands)	9,420	9,535	9,672
MDGs			
Access to improved water sources 2010 (%)[a]	87	87	87
Access to improved sanitation 2010 (%)[a]	100	100	100
IBNET sourced data			
Number of utilities reporting in IBNET sample	10	10	13
Population served (water), (thousands)	7,468	7,771	11,095
Size of the sample: Total population living in service area (water supply), (thousands)	8,097	8,234	12,368
Services coverage			
1.1 Water coverage (%)	92	94	90
2.1 Sewerage coverage (%)	36	34	24
Operational efficiency			
13.2 Electrical energy costs vs. operating costs (%) (share of energy cost as % of operational expenses)	40	40	35
6.1 Nonrevenue water (%)	29.00	29.00	36.00
6.2 Nonrevenue water (m³/km/day)	33	32	34
12.3 Staff W/1,000 W population served (W/1,000 W population served)	0.90	0.90	0.90
15.1 Continuity of service (hrs/day) (duration of water supply, hours)	21.00	21.00	19.20
Financial efficiency			
8.1 Water sold that is metered (%)	59	59	65
23.1 Collection period (days)	242	228	242
23.2 Collection ratio (%)	88	90	95
18.1 Average revenue W & WW (US$/m³ water sold)	0.12	0.13	0.10
11.1 Operational cost W & WW (US$/m³ water sold)	0.11	0.12	0.11
24.1 Operating cost coverage (ratio)	1.03	1.07	0.94
Production and consumption			
3.1 Water production (l/person/day)	351.00	342.00	304.00
4.1 Total water consumption (l/person/day)	251.00	244.00	195.00
4.7 Residential consumption (l/person/day)	122	122	99
Poverty and affordability			
19.1 Total revenues/service population/GNI (% GNI per capita) (average revenues)	1.15	1.02	0.55
19.2 Annual bill for households consuming 6 m³ of water/month (US$/yr)	9.81	10.51	10.55
21.1 Ratio of industrial to residential tariff (level of cross-subsidy)	3.96	3.41	3.91

a. UNICEF and WHO 2012.

IBNET Indicator/Country: República Bolivariana de Venezuela

Latest year available	2006
Surface area (km^2)	912,050
GNI per capita, Atlas method (current US$)	6,090
Total population (thousands)	27,031
Urban population (%)	92
Total urban population (thousands)	24,869
MDGs	
Access to improved water sources 2010 (%)[a]	—
Access to improved sanitation 2010 (%)[a]	—
IBNET sourced data	
Number of utilities reporting in IBNET sample	17
Population served (water), (thousands)	22,677
Size of the sample: Total population living in service area (water supply), (thousands)	25,149
Services coverage	
1.1 Water coverage (%)	90
2.1 Sewerage coverage (%)	74
Operational efficiency	
13.2 Electrical energy costs vs. operating costs (%) (share of energy cost as % of operational expenses)	3
6.1 Nonrevenue water (%)	62.00
6.2 Nonrevenue water (m^3/km/day)	138
12.3 Staff W/1,000 W population served (W/1,000 W population served)	0.60
15.1 Continuity of service (hrs/day) (duration of water supply, hours)	20.00
Financial efficiency	
8.1 Water sold that is metered (%)	38
23.1 Collection period (days)	416
23.2 Collection ratio (%)	91
18.1 Average revenue W & WW (US$/m^3 water sold)	0.25
11.1 Operational cost W & WW (US$/m^3 water sold)	0.26
24.1 Operating cost coverage (ratio)	0.95
Production and consumption	
3.1 Water production (l/person/day)	473.00
4.1 Total water consumption (l/person/day)	178.00
4.7 Residential consumption (l/person/day)	128
Poverty and affordability	
19.1 Total revenues/service population/GNI (% GNI per capita) (average revenues)	0.27
19.2 Annual bill for households consuming 6 m^3 of water/month (US$/yr)	—
21.1 Ratio of industrial to residential tariff (level of cross-subsidy)	—

a. UNICEF and WHO 2012.

IBNET Indicator/Country: Vietnam

Latest year available	2007	2008	2009
Surface area (km²)	331,212	331,212	331,212
GNI per capita, Atlas method (current US$)	790	920	1,120
Total population (thousands)	84,221	85,122	86,025
Urban population (%)	29	29	30
Total urban population (thousands)	24,025	24,812	25,610
MDGs			
Access to improved water sources 2010 (%)[a]	95	95	95
Access to improved sanitation 2010 (%)[a]	76	76	76
IBNET sourced data			
Number of utilities reporting in IBNET sample	69	66	66
Population served (water), (thousands)	16,339	18,096	19,613
Size of the sample: Total population living in service area (water supply), (thousands)	22,717	24,662	25,858
Services coverage			
1.1 Water coverage (%)	69	72	75
2.1 Sewerage coverage (%)	2	—	—
Operational efficiency			
13.2 Electrical energy costs vs. operating costs (%) (share of energy cost as % of operational expenses)	24	19	18
6.1 Nonrevenue water (%)	32.00	31.00	30.00
6.2 Nonrevenue water (m³/km/day)	42	43	40
12.3 Staff W/1,000 W population served (W/1,000 W population served)	1.20	1.10	1.10
15.1 Continuity of service (hrs/day) (duration of water supply, hours)	22.20	22.60	22.70
Financial efficiency			
8.1 Water sold that is metered (%)	100	100	100
23.1 Collection period (days)	250	263	245
23.2 Collection ratio (%)	98	—	—
18.1 Average revenue W & WW (US$/m³ water sold)	0.24	0.25	0.26
11.1 Operational cost W & WW (US$/m³ water sold)	0.12	0.15	0.15
24.1 Operating cost coverage (ratio)	2.05	1.72	1.68
Production and consumption			
3.1 Water production (l/person/day)	215.00	215.00	211.00
4.1 Total water consumption (l/person/day)	147.00	147.00	149.00
4.7 Residential consumption (l/person/day)	99	100	103
Poverty and affordability			
19.1 Total revenues/service population/GNI (% GNI per capita) (average revenues)	1.63	1.46	1.26
19.2 Annual bill for households consuming 6 m³ of water/month (US$/yr)	14.10	14.27	16.12
21.1 Ratio of industrial to residential tariff (level of cross-subsidy)	2.14	2.05	2.07

a. UNICEF and WHO 2012.

IBNET Indicator/Country: West Bank and Gaza

Latest year available	2008	2009	2010
Surface area (km²)	6,020	6,020	6,020
GNI per capita, Atlas method (current US$)	1,300	1,400	1,600
Total population (thousands)	3,937	3,920	3,905
Urban population (%)	73	74	74
Total urban population (thousands)	2,894	2,889	2,887
MDGs			
Access to improved water sources 2010 (%)[a]	—	—	—
Access to improved sanitation 2010 (%)[a]	—	—	—
IBNET sourced data			
Number of utilities reporting in IBNET sample	6	6	6
Population served (water), (thousands)	901	1,048	1,272
Size of the sample: Total population living in service area (water supply), (thousands)	914	1,071	1,315
Services coverage			
1.1 Water coverage (%)	99	98	97
2.1 Sewerage coverage (%)	56	57	58
Operational efficiency			
13.2 Electrical energy costs vs. operating costs (%) (share of energy cost as % of operational expenses)	25	19	20
6.1 Nonrevenue water (%)	44.00	45.00	44.00
6.2 Nonrevenue water (m³/km/day)	29	35	32
12.3 Staff W/1,000 W population served (W/1,000 W population served)	0.90	0.90	0.80
15.1 Continuity of service (hrs/day) (duration of water supply, hours)	16.00	16.40	14.70
Financial efficiency			
8.1 Water sold that is metered (%)	99	100	100
23.1 Collection period (days)	978	776	585
23.2 Collection ratio (%)	66	57	44
18.1 Average revenue W & WW (US$/m³ water sold)	1.12	0.89	1.31
11.1 Operational cost W & WW (US$/m³ water sold)	1.01	1.19	1.51
24.1 Operating cost coverage (ratio)	1.01	1.19	1.51
Production and consumption			
3.1 Water production (l/person/day)	115.00	133.00	123.00
4.1 Total water consumption (l/person/day)	64.00	80.00	73.00
4.7 Residential consumption (l/person/day)	57	80	69
Poverty and affordability			
19.1 Total revenues/service population/GNI (% GNI per capita) (average revenues)	2.01	1.86	2.18
19.2 Annual bill for households consuming 6 m³ of water/month (US$/yr)	96.33	83.40	101.11
21.1 Ratio of industrial to residential tariff (level of cross-subsidy)	1.65	0.86	11.53

a. UNICEF and WHO 2012.

IBNET Indicator/Country: Republic of Yemen

Latest year available	2008	2009	2010
Surface area (km²)	527,968	527,968	527,968
GNI per capita, Atlas method (current US$)	980	1,110	1,220
Total population (thousands)	22,627	23,328	24,053
Urban population (%)	31	31	32
Total urban population (thousands)	6,928	7,274	7,635
MDGs			
Access to improved water sources 2010 (%)[a]	55	55	55
Access to improved sanitation 2010 (%)[a]	53	53	53
IBNET sourced data			
Number of utilities reporting in IBNET sample	10	10	10
Population served (water), (thousands)	2,450	2,539	2,611
Size of the sample: Total population living in service area (water supply), (thousands)	3,405	3,638	3,756
Services coverage			
1.1 Water coverage (%)	63	61	61
2.1 Sewerage coverage (%)	61	61	64
Operational efficiency			
13.2 Electrical energy costs vs. operating costs (%) (share of energy cost as % of operational expenses)	17	24	24
6.1 Nonrevenue water (%)	33.00	34.00	33.00
6.2 Nonrevenue water (m³/km/day)	18	17	18
12.3 Staff W/1,000 W population served (W/1,000 W population served)	1.30	1.20	1.20
15.1 Continuity of service (hrs/day) (duration of water supply, hours)	16.40	16.20	16.80
Financial efficiency			
8.1 Water sold that is metered (%)	104	104	104
23.1 Collection period (days)	428	407	364
23.2 Collection ratio (%)	115	122	71
18.1 Average revenue W & WW (US$/m³ water sold)	0.49	0.57	0.56
11.1 Operational cost W & WW (US$/m³ water sold)	0.64	0.72	0.68
24.1 Operating cost coverage (ratio)	0.61	0.59	0.59
Production and consumption			
3.1 Water production (l/person/day)	108.00	105.00	106.00
4.1 Total water consumption (l/person/day)	74.00	71.00	73.00
4.7 Residential consumption (l/person/day)	58	57	59
Poverty and affordability			
19.1 Total revenues/service population/GNI (% GNI per capita) (average revenues)	1.35	1.33	1.22
19.2 Annual bill for households consuming 6 m³ of water/month (US$/yr)	18.16	22.71	24.19
21.1 Ratio of industrial to residential tariff (level of cross-subsidy)	2.84	3.24	3.18

a. UNICEF and WHO 2012.

IBNET Indicator/Country: Zambia

Latest year available	2011	2012	2013
Surface area (km^2)	1,160	1,160	1,160
GNI per capita, Atlas method (current US$)	1,180	1,350	1,400
Total population (thousands)	13,475	13,475	13,840
Urban population (%)	39	39	40
Total urban population (thousands)	5,255	5,255	5,481
MDGs			
Access to improved water sources 2010 (%)[a]	61	61	61
Access to improved sanitation 2010 (%)[a]	48	48	48
IBNET sourced data			
Number of utilities reporting in IBNET sample	11	11	11
Population served (water), (thousands)	4,243	4,435	4,626
Size of the sample: Total population living in service area (water supply), (thousands)	5,559	5,620	5,682
Services coverage			
1.1 Water coverage (%)	76	79	82
2.1 Sewerage coverage (%)	53	55	56
Operational efficiency			
13.2 Electrical energy costs vs. operating costs (%) (share of energy cost as % of operational expenses)	—	—	—
6.1 Nonrevenue water (%)	46.00	55.00	43.00
6.2 Nonrevenue water (m^3/km/day)	—	—	—
12.3 Staff W/1,000 W population served (W/1,000 W population served)	—	—	—
15.1 Continuity of service (hrs/day) (duration of water supply, hours)	16.30	16.30	17.10
Financial efficiency			
8.1 Water sold that is metered (%)	—	—	—
23.1 Collection period (days)	—	—	—
23.2 Collection ratio (%)	81	63	66
18.1 Average revenue W & WW (US$/m^3 water sold)	0.44	0.61	0.52
11.1 Operational cost W & WW (US$/m^3 water sold)	0.35	0.47	0.39
24.1 Operating cost coverage (ratio)	1.25	1.28	1.35
Production and consumption			
3.1 Water production (l/person/day)	227.00	217.00	197.00
4.1 Total water consumption (l/person/day)	123.00	98.00	112.00
4.7 Residential consumption (l/person/day)	77	73	74
Poverty and affordability			
19.1 Total revenues/service population/GNI (% GNI per capita) (average revenues)	1.67	1.62	1.52
19.2 Annual bill for households consuming 6 m^3 of water/month (US$/yr)	23.37	23.22	23.26
21.1 Ratio of industrial to residential tariff (level of cross-subsidy)	—	—	—

a. UNICEF and WHO 2012.

Reference

UNICEF and WHO (World Health Organization). 2012. *Progress on Drinking Water and Sanitation: 2012 Update.* www.unicef.org/media/files/JMPreport2012.pdf.

Index

Boxes, figures, notes, and tables are indicated by "b," "f," "n," and "t" following page numbers.

A

ACWUA (Arab Countries Water Utilities Association), xi

affordability of water and sanitation services, 3, 19–20, 19*t*

Africa. *See also specific regions and countries*
    staff productivity in, 12
    water consumption in, 18

Albania, data table for, 37*t*

Algeria, data table for, 38*t*

Apgar score for water and sanitation utilities, 25–29
    application of, 26–29, 27–28*f*, 28*t*, 33*n*1
    criteria, 25, 26–27*t*
    income levels and, 28–29, 28*t*
    overview, 4
    utility size and, 29, 29*f*
    value to utility managers and water authorities, 34*n*3

AquaRating system, 24*b*, 24*f*

Arab Countries Water Utilities Association (ACWUA), xi

Arab Republic of Egypt, data table for, 68*t*

Argentina, data table for, 39*t*

Armenia, data table for, 40*t*

Asia. *See specific regions and countries*

assessment. *See* measures of performance

Australia, data table for, 41*t*

Azerbaijan, data table for, 42*t*

B

Bahrain, data table for, 44*t*

Bangladesh, data table for, 43*t*

Belarus, data table for, 45*t*

benchmarking. *See* Apgar score for water and sanitation utilities; International Benchmarking Network for Water and Sanitation Utilities (IBNET)

Benin
    cost recovery vs. vulnerability in, 32, 33*f*
    data table for, 46*t*

Bhutan, data table for, 47*t*

billing and collections, 2, 18, 19*t*

Bolivia (Plurinational State of), data table for, 48*t*

Bosnia and Herzegovina, data table for, 49*t*

Brazil, data table for, 50*t*

Bulgaria, data table for, 51*t*

Burkina Faso, data table for, 52*t*

bursts and breaks of pipes, 3, 16*b*

Burundi, data table for, 53*t*

C

Cabo Verde, data table for, 54*t*

Cambodia, data table for, 55*t*

Cameroon, data table for, 56*t*

Caribbean. *See* Latin America and Caribbean

Central African Republic, data table for, 57*t*

Central Asia. *See* Eastern Europe and Central Asia

Chile, data table for, 58*t*

China, data table for, 59*t*

collection of data. *See* data collection and analysis

collections. *See* billing and collections

Colombia, data table for, 60*t*

Congo (Democratic Republic of), data table for, 61*t*

consumption of water, 3, 17–18, 18*t*

Costa Rica, data table for, 62*t*

cost recovery ratio (CRR), 2, 2*f*

Côte d'Ivoire, data table for, 63*t*

country data tables, 37–138. *See also specific countries*

Croatia, data table for, 64*t*

ECO-AUDIT
Environmental Benefits Statement

The World Bank is committed to preserving endangered forests and natural resources. The Office of the Publisher has chosen to print *The IBNET Water Supply and Sanitation Blue Book* on recycled paper with 100 percent postconsumer fiber in accordance with the recommended standards for paper usage set by the Green Press Initiative, a nonprofit program supporting publishers in using fiber that is not sourced from endangered forests. For more information, visit www .greenpressinitiative.org.

Saved:
- 9 trees
- 4 million Btu of total energy
- 807 lb. of net greenhouse gases
- 4,381 gal. of waste water
- 293 lb. of solid waste

green
press
INITIATIVE

www.ingramcontent.com/pod-product-compliance
Lightning Source LLC
Chambersburg PA
CBHW082356270326
41935CB00013B/1647